Sericulture Manual

Sericulture Manual

by

R.K. Patnaik

BIOTECH

2025

Biotech Books

ISBN10 81-7622-188-0
ISBN13 978-81-7622-188-7
EAN: 9788176221887

Published by : **BIOTECH BOOKS**
4762-63/23, Ansari Road, Darya Ganj,
New Delhi - 110002
Phones : 011 4300 3222
Email : biotechbooks@yahoo.co.in

PRINTED IN INDIA

Preface

When the Ancient Chinese discovered the art of silk culture in the 3rd millennium BC, little idea did they have of the large scale industry they had initiated. The fibre which was once considered the prerogative of the Chinese nobility, has metamorphosed into a major economic product generating millions in revenues for numerous countries; and has become almost as common as cotton, albeit more expensive.

Sericulture refers to the production of raw silk. Bombyx mori is the most widely studied and used commercial variety of silkworm species, and per Chinese records, they were begun to be used around 2700 B.C. Today, China and Japan are two main producers of silk, and account for more than 50% of the total silk production.

This book has been brought out to facilitate the reader's knowledge about the art of rearing silkworks for silk. The Chinese kept the knowledge of sericulture a closely guarded secret, but sericulture today is a widely practised out, employing the latest technology to aid in its tedious process. This book covers a wide gamut of the sericulture arena - beginning from the feeding of mulberry leaves to the silkworms, right upto silk finishing and its trade. Comprehensive chapters detailing each step of the sericulture process from the crux of the book, and illustrations further enrich the learning process. Also, the silk industry, its history and its future, and its size and scope, have also been given a thought to, and find themselves included in the manual.

Providing the reader with a thoroughly explored compendium on the sericulture industry, it is hoped that this book enlightens the readers.

Editor

Contents

1

Mulberry Cultivation

Mulberry is a family of deciduous or evergreen trees and shrubs, often climbing, mostly of pantropical distribution, and characterised by milky sap. Several genera bear edible fruit, e.g., *Morus,* (true mulberries), *Ficus* (the fig genus), and *Artocarpus,* which includes the breadfruit and related species. The related hemp family, whose plants do not contain latex, were formerly included in this family.

Mulberry (Morus) is a genus of 10–16 species of deciduous trees native to warm temperate and subtropical regions of Asia, Africa and North America, with the majority of the species native to Asia. The closely related genus Broussonetia is also commonly known as mulberry, notably the Paper Mulberry Broussonetia papyrifera.

Mulberries are fast-growing when young, but soon become slow-growing and rarely exceed 10-15 m tall. The leaves are alternately arranged, simple, often lobed, more often lobed on juvenile shoots than on mature trees, and serrated on the margin. The fruit is a multiple fruit, 2-3 cm long, red ripening to dark purple, edible, and sweet with a good flavour in several species.

The taxonomy of Morus is complex and disputed. Over 150 species names have been published, but only

10-16 are generally cited as being accepted, though different sources cite different selections of accepted names. The classification is also complicated by widespread hybridisation, with the hybrids being fertile. The following species are generally accepted:

— *Morus alba* (White Mulberry; eastern Asia)

— *Morus australis* (Chinese Mulberry; southeastern Asia)

— *Morus mesozygia* (African Mulberry; southern and central Africa)

— *Morus microphylla* (Texas Mulberry; south central North America: Texas, Mexico)

— *Morus nigra* (Black Mulberry; southwest Asia)

— *Morus rubra* (Red Mulberry; eastern North America)

The following, all from eastern and southern Asia, are additionally accepted by one or more taxonomic lists or studies:

— *Morus atropurpurea*

— *Morus bombycis*

— *Morus cathayana*

— *Morus indica*

— *Morus japonica* (*M. alba*)

— *Morus kagayamae* (*M. australis*)

— *Morus laevigata* (*M. alba* var. *laevigata*, *M. macroura*)

— *Morus latifolia* (*M. alba*)

— *Morus liboensis*

— *Morus macroura* (*M. alba* var. *laevigata*)

— *Morus mongolica* (*M. alba* var. *mongolica*)

— *Morus multicaulis* (*M. alba*)

— *Morus notabilis*

— *Morus rotundiloba*

— *Morus serrata* (Himalayan Mulberry; *M. alba* var. *serrata*)

— *Morus tillaefolia*

— *Morus trilobata* (*M. australis* var. *trilobata*)

— *Morus wittiorum*

The mulberry family is most important as the basis of the silkworm industry; silkworms feed on the leaves of the mulberries (genus *Morus*) and sometimes of the Osage orange (*Maclura pomifera*). The white mulberry (*M. alba*) has been cultivated in China since very early times. In the Middle Ages it began to replace the black mulberry (*M. nigra*), which had been grown by the Greeks and Romans and, from the 9th cent., by the people of N Europe for silkworm culture.

In Greek legend the berries of the white mulberry turned red when its roots were bathed by the blood of the lovers Pyramis and Thisbe, who killed themselves. Both the white and the red mulberry (*M. rubra,* native to North America) have been cultivated in America since colonial times, but the lack of cheap hand labor prevented the establishment of a silkworm industry. Mulberry fruits are tender and juicy and resemble blackberries. In the South the fruit of *M. rubra* is made into wine and is considered a valuable agricultural and wildlife feed.

The Osage orange, also called bowwood because it was used by the Osage tribe to make bows, is a hardy tree native to the S central United States. Its fruit is used as a natural insect repellent. Cultivated widely, often as a hedge plant because of its spiny, impenetrable branches, it is a source of a flexible and durable wood and of a yellow-orange dye, from the root bark, that is similar to the more widely used fustic (*Maclura tinctoria*). The heartwood of fustic yields a yellowish or olive dye, also called fustic, that has been used chiefly for dyeing woolens; it has largely been replaced by synthetic aniline dyes.

In its native habitat of Central and South America the fustic is also a timber tree. Fiber plants of the mulberry family include the paper mulberry (*Broussonetia papyrifera*) and the upas tree (*Antiaris toxicara*) of the East Asian tropics, where the bast fiber is utilised for rough fabrics and for paper, often after a crude retting process. The latex of the upas [Malay,=poison tree] contains a cardiac glycoside used for arrow poison; the similarly employed strychnine tree of the logania family is sometimes also called upas.

The breadfruit (*Artocarpus ultilis*) is cultivated as a staple food plant in the Pacific tropics and in the West Indies, where it was introduced from Polynesia in the late 18th cent.; the *Bounty* was carrying breadfruit plants to Jamaica when the famous mutiny occurred. Its wood, fiber, and latex are also variously utilised locally. The important fig genus includes fruit trees, ornamentals (e.g., the rubber plant), and several species renowned in the religion and legends of India (e.g., the banyan and the bo tree).

Uses and Cultivation of Mulberry

The fruit is edible and is widely used in pies, tarts, wines and cordials. The fruit of the Black Mulberry, native to southwest Asia, and the Red Mulberry, native to eastern North America, have the strongest flavour. The fruit of the White Mulberry, an east Asian species which is extensively naturalised in urban regions of eastern North America has a different flavour, sometimes characterised as insipid. The mature fruit contains significant amounts of resveratrol. Mulberry leaves, particularly those of the White Mulberry, are also economically important as the sole food source of the silkworm, the cocoon of which is used to make silk.

Other Lepidoptera larvae also sometimes feed on the plant including Common Emerald, Lime Hawk-moth and The Sycamore. Mulberries can be grown from seed, and this is often advised as seedling-grown trees are generally of better shape and health. However, they are most often planted from large cuttings, which take root readily.

Types of Mulberries

White Mulberry (*Morus alba* L.), Black Mulberry (*M. nigra* L.), American Mulberry and Red Mulberry (*M. rubra* L.) are the different types of mulberries

The white mulberry is native to eastern and central China. It became naturalised in Europe centuries ago. The tree was introduced into America for silkworm culture in early colonial times and naturalised and hybridised with the native red mulberry. The red or American mulberry is native to eastern United States from Massachusetts to Kansas and down to the Gulf coast. The black mulberry is native to western Asia and has been grown for its fruits in Europe since before Roman times.

The white mulberry, and to a lesser extent the red mulberry, are quite tolerant of drought, pollution and poor soil. The white mulberry is considered a weed tree in many parts of the country including urban areas. The black mulberry is more fastidious, faring less well in cold climates or areas with humid summers. The white mulberry is the most cold-hardy of the three species, although this varies from one clone to another. Some are damaged at 25° F, while others are unfazed at -25° F.

Red mulberries are hardy to sub-zero temperatures. The black mulberry is the least cold-hardy of the three, although again cold tolerance seems to depend on the clone. In general it is limited to USDA Hardiness Zone 7 (0° to 10° F average minimum) or warmer. They have been planted only to a limited extent in America, mostly on the Pacific Coast. The mulberry makes a good town tree which will grow well in a tub.

All three mulberry species are deciduous trees of varying sizes. White mulberries can grow to 80 ft. and are the most variable in form, including drooping and pyramidal shapes. In the South on rich soils the red mulberry can reach 70 ft. in height. The black mulberry is the smallest of the three, sometimes growing to 30 ft. in height, but it tends to be a bush if not trained when it is young. The species vary greatly in longevity. Red mulberry trees rarely live more than 75 years, while black mulberries have been known to bear fruit for hundreds of years. The mulberry makes an attractive tree which will bear fruit while still small and young.

The white mulberry is so-named for the color of its buds, rather than the color of its fruit. The thin, glossy, light green leaves are variously lobed even on the same plant. Some are unlobed while others are glove-shaped. Leaves of the red mulberry are larger and thicker, blunt toothed and often lobed. They are rough on their upper surfaces and pubescent underneath. The smaller black mulberry leaves are similar to those of the red mulberry, but with sturdier twigs and fatter buds. The species vary in the time of year they begin to leaf-out. White mulberries generally come out in early spring, almost two months before black mulberries.

Mulberry trees are either dioecious or monoecious, and sometimes will change from one sex to another. The flowers are held on short, green, pendulous, nondescript catkins that appear in the axils of the current season's growth and on spurs on older wood. They are wind pollinated and some cultivars will set fruit without any pollination. Cross-pollination is not necessary. In California mulberries set fruit without pollination.

Botanically the fruit is not a berry but a collective fruit, in appearance like a swollen loganberry. When the flowers are pollinated, they and their fleshy bases begin to swell. Ultimately they become completely altered in texture and color, becoming succulent, fat and full of juice. In appearance, each tiny swollen flower roughly resembles the individual drupe of a blackberry. The color of the fruit does not identify the mulberry species. White mulberries, for example, can produce white, lavender or black fruit.

White mulberry fruits are generally very sweet but often lacking in needed tartness. Red mulberry fruits are usually deep red, almost black, and in the best clones have a flavor that almost equals that of the black mulberry. Black mulberry fruits are large and juicy, with a good balance of sweetness and tartness that makes them the best flavored species of mulberry. The refreshing tart taste is in some ways reminiscent of grapefruit. Mulberries ripen over an extended period of time unlike many other fruits which seem to come all at once.

Culture

Location: Mulberries need full sun and also adequate space. The distance between trees should be at least 15 ft. The trees should not be planted near a sidewalk. The fallen fruit will not only stain the walkway, but are likely to be tracked indoors. The trees are quite wind-resistant with some cultivars used as windbreaks in the Great Plains region.

Soil: Mulberries like a warm, well-drained soil, preferably a deep loam. Shallow soils such as those frequently found on chalk or gravel are not recommended.

Irrigation: Although somewhat drought-resistant, mulberries need to be watered in dry seasons. If the roots become too dry during drought, the fruit is likely to drop before it has fully ripened.

Fertilisation: Mulberries generally thrive with minimal fertilisation. An annual application of a balanced fertiliser such as 10:10:10 NPK will maintain satisfactory growth. In California mulberries usually need only nitrogen.

Pruning: No special pruning techniques are needed after the branches have been trained to a sturdy framework, except to remove dead or overcrowded wood. A mulberry tree can be kept to a tidy form by developing a set of main branches, and then pruning laterals to 6 leaves in July in order to develop spurs near the main branches. It is not advisable to prune the trees heavily since the plant is inclined to bleed at the cuts. Cuts of more than two inches in diameter generally do not heal and should be avoided at all cost. The bleeding will be less severe if the tree is pruned while it is dormant.

Propagation: Mulberries can be grown from seed, although the plants can take 10 years or more to bear. Seed should be sown as soon as extracted from the fruit, although white mulberry seeds germinate better after stratifying one to three months before planting.

Sprig budding is the most common method for grafting mulberries. A T-cut is made in the rootstock and a smooth, sloping cut is made on the lower end of the scion. The scion is then inserted into the T and wrapped and sealed. Other types of grafts are also usually successful, although there may be incompatibility between white and black mulberries. Hardwood, softwood and root cuttings also are suitable methods for propagating mulberries. Softwood cuttings of white mulberries root easily when taken in midsummer and treated with rooting hormone. Red mulberries are less easily rooted. Black mulberries are also somewhat difficult to propagate since they tend to bleed a lot.

Pests and Diseases: Mulberries are generally free of pests and diseases, although cankers and dieback can occur. In some areas "popcorn disease" is an occasional problem, in which fruits swell to resemble popped corn. *M. alba/M. rubra* hybrids are particularly prone to this condition. The disease carries on from one season to the next, so collecting and burning infected fruits help control it. The ripe fruit is very attractive to birds, but there is usually enough fruit left over for harvesting.

Harvest: White and red mulberry fruits (and hybrid fruits) are ready for harvest in late spring. The fruit of black mulberries ripen in summer to late summer. The fruits of white mulberries are often harvested by spreading a sheet on the ground and shaking the limbs. A surprising quantity can be gathered from a comparatively small and young tree. Black mulberry fruits are more difficult to pick.

As the berries are squeezed to pull them loose, they tend to collapse, staining the hands (and clothing) with blood red juice. Unwashed the berries will keep several days in a refrigerator in a covered container. The ripe fruits of the black mulberry contain about 9% sugar with malic and citric acid. The berries can be eaten out of hand or used in any way that other berries are used, such as in pies, tarts, puddings or sweetened and pureed as a sauce. Slightly unripe fruits are best for making pies and tarts. Mulberries blend well with other fruits, especially pears and apples. They can also be made into wine and make an excellent dried fruit, especially the black varieties.

Cultivars

Black Persian: *M. nigra.* Large black fruit, over an inch long and almost as wide. Juicy with a rich, subacid flavor. The tree is fairly drought-resistant once established.

Collier: *M. alba* X *M. rubra.* Medium-sized, purplish-black fruit, 1-1/8 inches long and 3/8 inch in diameter. Flavor sweet, with just a trace of tartness. Quality very good, on par with Illinois Everbearing. Ripens over a long period. Tree of medium size, spreading, relatively hardy, very productive.

Downing: The original Downing was a *M. alba* var. *multicaulis* plant grown fromseed sown about 1846. The fruit was black with excellent flavor and ripened from June to September. Other varieties have subsequently been sold under the same name.

Illinois Everbearing: *M. alba* X *M. rubra.* Originated in White County, Illinois. Introduced in 1958. Black, nearly seedless fruit large and very long, averaging 12 per ounce. Flavor good to very good, very sweet, considered best by by many. Matures over along season. Tree vigorous and somewhat dwarfed, extremely hardy and productive.

Kaester: *M. nigra.* Originated in Los Angeles. Introduced in 1971 by Nelson Westree. Large black or deep purple, elongated fruit, 1-1/2 inches long and 1/2 inch in diameter. Flavor very sweet, with good sweet/tart balance. Tree bears heavily.

Pakistan: Originated in Islamabad, Pakistan. Extremely large ruby-red fruit 2-1/2 to 3-1/2 inches long and 3/8 inch in diameter. Flesh firmer than most other named cultivars. Sweet with a fine balance of flavors. Quality excellent. Tree spreading with large heart-shaped leaves. Recommended for the deep South and mild winter areas such as southern California, but usually performs satisfactorily in cooler areas.

Riviera: Originated in Vista, Calif. Elongated, deep purple-black fruits, 1 to 1-1/2 inches in length, 1/2 inch in diameter. Flesh slightly juicy and very sweet. Very good dessert quality. Ripens over a long period, from April to June.

Russian (Tatarica): Introduced into Europe from China about 1,500 years ago. Fruit reddish-black, of good quality when completely ripe. Tree bushy, to 35 ft. tall, very hardy and drought resistant. Planted widely for windbreaks and wildlife food.

Shangri-La: Originated in Naples, Fla. Large, black fruit. Good mulberry for the Deep South and other areas. Hardy in U.S.D.A. Zones 7-9. Tree has very large, heart-shaped leaves.

Tehama (Giant White): Originated in Tehama County, Calif. Very large, white-colored, plump fruit, 2-3/4 inches in length and 1/2 inch wide. Very sweet, succulent, melting flesh. Attractive, large-leaved tree. Probably best adapted to mild winter areas.

Wellington: Originated in Geneva, N.Y. Reddish-black medium-sized fruit, 1-1/4 inches long, 3/8 inch in diameter. Form long, slender and cylindrical. Flesh soft, of good flavor. Ripens over a period of several weeks. Tree is heavy producer. May be the old cultivar New American, which was also sold many years ago as Downing.

Uses

Trees are extensively grown (e.g. southern Europe, India) for their leaves as food for silkworms. Fruits may be eaten raw or cooked. Fruits are an ingredient of a particularly seductive drink known as Mulberry Wine. Wild birds, poultry, and hogs are fond of fruits. Mulberry leaves are sometimes eaten as a vegetable and are useful as a cattle fodder. Being nutritious and palatable, they are said to improve milk yield of dairy animals. Wood is valued for sporting goods due to its elasticity and flexibility when steamed, being considered as good as ash.

Sapwood is white to yellowish-white, heart-wood golden brown, darkening after exposure; light to moderately heavy, straight-grained, not liable to split, but shows tendency to warp, moderately durable under cover, easy to work and finish, used mainly for hockey sticks, tennis and badminton rackets and racket presses, cricket bats, house building materials, agricultural implements, and furniture. Furnishes a medium grade fuel wood. Stem bark is fibrous and used in China and Europe for paper making. Twigs are used as binding material and for making baskets. Trees often used as ornamentals, roadsides or boundary markers.

Folk Medicine

The fruit juice is used in folk remedies for tumors of the fauces. Reported to be antidotal, antiphlogistic, antitussive. antivinous, astringent, bactericide, diaphoretic, ditiretic, emollient, escharotic, expectorant, fungicide, laxative, nervine, purgative, refrigerant, restorative, sedative, tonic, and vermifuge, white mulberry is a folk remedy for aphtha, armache, asthma, bronchitis, bugbite, cachexia, cold, constipation, cough, debility, diarrhea, dropsy, dyspepsia, edema, epilepsy, fever, headache, hyperglycemia, hypertension, inflammation, insomnia, melancholy, menorrhagia, snakebite, sorethroat, stomatitis, tumors, vertigo, and wounds.

Medicinally, fruits are laxative, refrigerant in fevers, and used locally as remedy for sore throat, dyspepsia, and melancholia. Roots and bark are purgative, anthelmintic, and astringent; leaves considered disphoretic and emollient; a decoction of leaves being used as a gargle for inflammation of throat.

Per 100 g, the fruit is reported to contain 87.5 g H_2O, 1.5 g protein, 0.49 g fat, 8.3 g carbohydrates, 1.4 g fiber, 0.9 g ash, 80 mg Ca, 40 mg P, 1.9 mg Fe, 174 IU vit. A, 9 mg thiamine, 184 mg riboflavin, 0.8 mg nicotinic acid, and 13 mg ascorbic acid. Dry leaves contain 18–28.8% protein, 11.3–0.7% MgO, 0.8–13.6% soluble sugars, 0.6–1.4% P_2O_5, 2–3.9% K_2O, 1.4–2.4% CaO, 0.2–0.7% MgO, 0.8–1.8% Al_2O_3, 0.05–0.26% Fe_2O_3, 1.8–2.6% SiO_2, and 0.3–0.56% SO_4 (C.S.I.R., 1948–1976).

Substances which attract the silkworm larvae to the leaves have been identified as citral, linalyl acetate, linalol, terpinyl acetate, and hexenol, the first 3 being more effective. b-sitosterol (ca 0.2% in leaves), along with some sterols and a water-soluble substance, is the main factor which stimulates the biting action; the amount of food eaten by larvae is controlled by the concentration of b-sitosterol. Protein perparations from young mulberry leaves form an excellent supplement to protein-deficient diets. Non-protein nitrogen accounts for ca 22% of the total nitrogen in young leaves and ca 14% in mature leaves.

Amino acids identified in the free form are: phenylalanine, leucine, valine, tyrosine, proline, alanine, glutamic acid, glycine, serine, arginine, aspartic acid, cystine, threonine, sarcosine, gamma-amino-butyric acid, pipecolic acid, and 5-hydroxy pipecolic acid. The leaves are a good source of ascorbic acid, with 2–3 mg/g, of which over 90% is present in the reduced form. They contain also carotene, vitamin B1, folic acid, folinic acid, and vitamin D. Phytate phosphorus accounts for 18.2% of total phosphorus.

Volatile constituents identified in steam-distillates of the leaves are: n-butanol, beta-gamma-hexenol, methyl-ethyl acetaldehyde, n-butylaldehyde, isobutylaldehyde, valeraldehyde, hexaldehyde, alpha-beta-hexenal, acetone, methyl-ethyl ketone, methyl-hexyl ketone, butylamine, and acetic, propionic, and isobutyric acids. Leaves also contain calcium malate, succinic, and tartaric acids, xanthophyll and isoquercitrin (quercetin 3-glucoside) and tannins; adenine, chorine, and trigonelline bases are present in young leaves. Analysis of leaves gave (dry weight basis): protein, 14.0; EE, 6.8; NFE, 49.7; total ash, 13.8; calcium (CaO, 2.74; and phosphorus (P_2O_5) 0.45% (C.S.I.R., 1948–1976).

According to Hager's Handbook, the wood contains 0.3–0.44% morin, dihydromorin, dihydrokaempferol, 2,4,4',6-tetra-hydroxybenzophenone, maclurin, and ca 2% hydroxyresveratrol. Seeds contain 33–38% fatty oil (with linoleic-, stearic-, oleic-, myristic-, palmitoleic-, and arachidic-acids) and ca 25% protein.

Small to medium-sized monoecious or dioecious shrub or tree, up to 15 m tall, wide-spreading, round-topped, trunk attaining 60 cm in diameter; leaves alternate, stipulate, variable in shape, lobed or unlobed, cordate, dentate, acuminate, long-petiolate, 12 x 8 cm on fruiting branches, up to 25 x 20 cm on vigorous non-fruiting branches, usually smooth above, glabrous or pubescent along veins beneath, thin, light green; flowers small, greenish-yellow, in dense spikes to 2 cm long; sepals 4; stamens 4; pistils with two styles; staminate spikes soon deciduous; pistillate spikes maturing into an aggregate fruit (syncarp) of drupelets; syncarp ovoid to oblong-cylindric, 1–5 cm long, white, pinkish or purplish to nearly black, edible long before ripe, sweet, but insipid; seeds brown, 1–1.2 mm long, 12,000–13,000/oz. Fl. Mar.–May, fr. Summer.

Germplasm

Reported from the China-Japan Center of Diversity, white mulberry, or cvs thereof, is reported to tolerate disease, drought, frost, hydrogen fluoride, low pH, poor soil, salt, shade, slope, and weeds. It is also said to tolerate aerial sprays of 2,4,5-T but not ground treatment. Numerous cvs have been adapted in various silk-producing regions of the world, differing in their adaptability to various soils and climates (as those suited for cold climate, for areas with severe winters, or for mild climate; some are early, others late or intermediate); resistance to diseases; food value of leaf for silkworms; and suitability for use as stock or scion in grafting.

Japan has about 700 types of which 21 are extensively cultivated. Most important cv grown in India is *M. alba* var. *muticaulis* Loud., fast growing, adapted to field culture, and giving high yield of large, tender, thick leaves. In India, mulberry is either grown as a field crop (plants kept in form of bushes and the leaves harvested several times a year for the multivoltine race of silkworms) or as a tree (leaves harvested only once a season for rearing univoltine races of silkworms). In Bengal, cultivation of the var. *atropurpurea* along borders of bush plantations has been recommended. Largest acreage of mulberry in India is in Mysore, where more than 75% of total mulberry raw silk production occurs.

There has been some selection of mulberry in India, the selection 'K.M.', in West Bengal, gives 50% more leaves than local types, and is popular in Mysore for grafting purposes. Also strains 'Selected I' and 'Selected V' evolved in Mysore are promising. Forty-four cvs, some local, others introduced, have been studied in Thailand. The Russian cvs (*M. alba* var. *tatarica*) are often planted as windbreaks, as they are very hardy and thrive under greatest neglect. For this purpose, they are planted 2.6–7 m apart, making excellent screens and standing pruning well.

Used as rootstocks for other mulberries, they are very popular in the Great Lakes Region of the United States. Raised from seed and quite variable, the fruits are inferior, suitable only for birds. 'New American', considered the best cv, 'Trowbridge' and 'Thorburn' are the leading kinds of fruit-bearing mulberries for the North. ($2n = 28$)

Native of China, white mulberry is cultivated throughout the world wherever silkworms are raised, and is occasionally cultivated elsewhere in Europe, North America, and Africa. Having escaped, trees often appear on roadsides, along fencerows, and as ornamentals.

Ecology

Ranging from Cool Temperate Steppe to Warm through Tropical Very Dry to Moist Forest Life Zones, white mulberry is reported to tolerate annual precipitation of 4.4 to 40.3 dm (mean of 28 cases = 9.8), annual temperature of 5.9 to 27.5°C (mean of 28 cases = 16.6), and pH of 4.9 to 8.0 (mean of 23 cases = 6.7). According to Wyman (1974), it tolerates average annual minimum temperatures of -10 to -20°F (-23.3 to -28.9°C). White mulberry grows well on wide variety of soils. It withstands drought, once well established. However, it is liable to wind damage. In India, it is cultivated up to 3,300 m elevation, but can grow as well at sea-level.

Cultivation

Mulberry trees can be propagated by seeds, cuttings, or graftings. Seeds should be treated with camphor water before sowing to ward off disease. Thin layer of soil and ashes spread over seed after sowing. Beds kept moist. Seeds germinate in 9–14 days, depending on the season. When seedlings are about 7.5 cm tall, they are thinned and weeded. For bush mulberries, seedlings 10–15 cm tall are used as transplants; for trees, seedlings are allowed to grow 1.3 m and trained before transplanting. Cuttings are obtained from exhausted bushes at pruning time.

Branches are cut into pieces 22–30 cm long with 3 buds and planted immediately. Sometimes cuttings are tied into bundles and kept buried in soil 6–9 cm deep, kept watered, and after about a month when buds have shoots about 5 cm long, are taken out of mud and planted. Cuttings root quickly, and with cultivation and irrigation, they may attain a height of 75 cm in 6 weeks. Mulberry plants from seedlings are more expensive, but give better plants than those from cuttings. Seedlings usually planted 2 to a pit 4" pits spaced 60–75 cm apart. Root grafting is usually practiced in India.

Stocks are seedlings of any type; the scion are selected from high yielding, quick-growing types with large leaves of good food value. Rooted cuttings are planted in pits or furrows. In dryland culture, pit system is uged. Cuttings placed in prepared pits, 30 cm in diameter and 22 cm deep, spaced 75 cm apart, in rows 75 cm apart. Planted in July after SW monsoon has set in. Usually 3 cuttings are put in each pit and watered; 75,000 to 120,000 cuttings/ha. When irrigation is used, cuttings are planted in furrows in April or May, 10 cm apart, the furrows being 22 cm apart. With this very close planting, 110,000 to 200,000 cuttings/ha are required.

Grafted plants develop a better root system than those from either seedlings, cuttings, or layerings, and are used exclusively in Japan. Grafted trees are planted 1.6 m apart each way, about 4,000/ha, and are especially suitable for irrigated areas. Various techniques are employed to prune and train mulberry plants. After each pruning, the field is cultivated and manured.

Harvesting

Fruits are picked or shaken on drop cloth's when in season. Not being attractive for marketing, they are usually consumed locally. For harvesting leaves, mulberry bushes under irrigation attain a height of about 1.3 m in 10 weeks, when the leaves are ready for picking. Under rainfed conditions, first picking may be done 12–17 weeks after planting. About 10 pickings per season for irrigated plants and 6 or 7 for rainfed crops. Best time for picking leaves is in the evening.

Harvested leaves are stored in small loose heaps in cool room and protected from heating, fermentation, or drying out. Bushes may be productive for as many as 15 years, after which old plants are pulled out, the land planted to a green manure crop and then replanted to mulberry bushes.

Yield of mulberry leaves varies according to the soil, moisture supply, manuring, and cultural practices. Bushes grown under rainfed conditions yield 4,000–7,000 kg/ha annually; under irrigation, from 10,000–14,000 kg/ha and selected varieties, as 'Selected I' and 'Selected V' yield as high as 22,000 kg/ha. Yields of leaves from tree mulberries are not available. Different trees vary greatly in fruit production. Sericulture is the most important commercial use for white mulberry. Young, fully developed leaves are best for feeding silkworm larvae.

In Thailand, local silkworms produce 300–800 m of fiber per cocoon compared with 1,600 m for Japanese silkworms, and 1,400 m for a cross between them. Those of us who have seen solid coats on the ground of white fruits of the white mulberry or red to purple fruits of other species, realise that there is a lot of sugar there that could be converted to alcohol.

Finding rather high yields of ethanol in peracetic acid treatments and four days incubation of the sawdust, Lee commented, "the yield of ethanol from wood sawdust can be increased from 1% to 6% if the treatment for delignification is improved. It is concluded that *Populus* species and *Morus* show great potential for ethanol production. We have established about 300,000 ha of improved, fast growing Populus hybrids (mostly *P.* x *euramericana* and *P. alba* x *glandulosa*) throughout the country. In Korea...a 45,000 ha plantation has been established with mulberry trees mainly to produce leaves for silkworms...

Most of the stems and branches produced by mulberry trees so far are utilised for heating or cooking and as a supporting material for horticultural crops.

Recently, preliminary studies on yields of ethanol from mulberry trees show promising possibilities. Also, if the conversion of mulberry tree biomass to fuels is economically feasible, planting mulberry trees will be greatly advantageous to Korean farmers. Other fuel sources might be organic wastes, mill residues, logging residues and agricultural by-products such as rice, barley, or wheat straw. Remarkable yields of energy may be expected on the assumption that the biomass produced from all existing plantations is converted to fuels. With 10% sulfuric acid sawdust hydrolysed for one hour yields 28% sugar.

White Mulberry Leaves and Fruit

On young, vigorous shoots, White Mulberry leaves may be up to 20 cm long, and deeply and intricately lobed, with the lobes rounded. On older trees, the leaves are generally 8-15 cm long, entire, cordate at the base and acuminate at the tip, and serrated on the margin. The fruit is mild, unlike the much more intense flavour of the Red Mulberry and Black Mulberry. The fruit varies from white to pink in colour in many cultivated plants, but the natural fruit colour of the species in the wild is deep purple.

The White Mulberry is scientifically notable for its rapid plant movement. The flowers fire pollen into the air by rapidly (25 μs) releasing stored elastic energy in the stamen. The resulting movement is in excess of half the speed of sound, making it the fastest movement in the plant kingdom.

The leaves are the preferred feedstock for silkworms, and are also cut for food for livestock (cattle, goats, etc.) in areas where dry seasons restrict the availability of ground vegetation. White Mulberry is extensively planted throughout the warm temperate Northern Hemisphere, and is naturalised in urban areas of the United States, where it hybridises to some extent with the U.S. native Red Mulberry (*Morus rubra*). In fact, some authorities worry about the long-term genetic viability of Red Mulberry because of extensive hybridisation in some areas. A weeping cultivar of White Mulberry *Morus alba* 'Pendula' is a popular ornamental plant. Weeping plants are propagated by grafting the weeping cultivar onto a non weeping rootstock.

2

Rearing of Silkworm

For silkworm rearing, usually eggs of white hybrids are used for spring, summer and fall rearing. White hybrids have high output of cocoons and silk. They also have high output of cocoons and silk. These hybrid silkworms present the following appreciable characteristics: a) high production of fresh cocoons (up to 30 Kg per box of 20.000 eggs), b) high output in silk (from 5.5 – 6 Kg of fresh cocoons we take 1Kg of silk), c) high resistance to diseases. In addition, their rearing is completed in short time (28 - 30 days), which considerably decreases the cost of production.

Rearing silkworms for only 30 days may ensure a very good income for the producer, particularly during a period where, in most of the cases, there is no other income from other agricultural products.

Figure 1a. Silkworm eggs that just oviposited

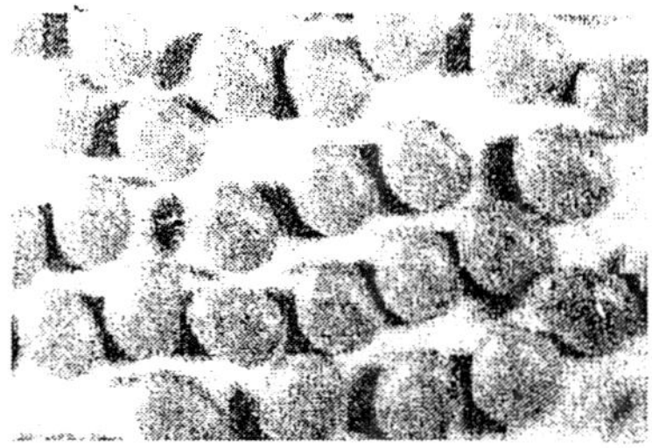

Figure 1b. Silkworm eggs after several days

Incubation of Eggs

After being laid, eggs are of yellow colour and several hours later, they become darker and turn gray which is their final colour (Fig. 1). The unfertilized eggs remain yellow. Overwintering eggs are kept at low temperature (2.5°C) till the next spring and until the beginning of the incubation process. Egg hatching very much depends on climatic parameters such as temperature, relative humidity and light intensity. These parameters are of vital importance since they can influence the incubation period, the development of silkworm embryo and ultimately the cocoon production.

The incubation takes place in special places (incubating chambers) where suitable conditions of temperature, humidity, light and air are created. If the incubation does not take place in an incubating chamber, then it is recommended to carry out it in a specially adapted place (e.g. small room) where the desirable conditions must be maintained. It is recommended to begin the incubation with the emergence of the first mulberry leaves. If the silkworms emerge earlier they will not have food; while if they emerge later they will not develop correctly because at this time the leaves will be very hard. The hatching of larvae should be synchronous (in 2-3 days) at least in 95 % of cases. Only the eggshells remain after the completion of the egg hatching (Fig. 2).

Incubation Conditions

Temperature

Temperature is considered to be the most important factor for silkworm egg hatching. It affects the duration of incubation, uniformity and percentage of hatching, quality of the cocoons and the voltinism of silkworm. The effect of temperature conditions on egg hatching depends also on whether the hatching is achieved naturally or artificially by acid treatment.

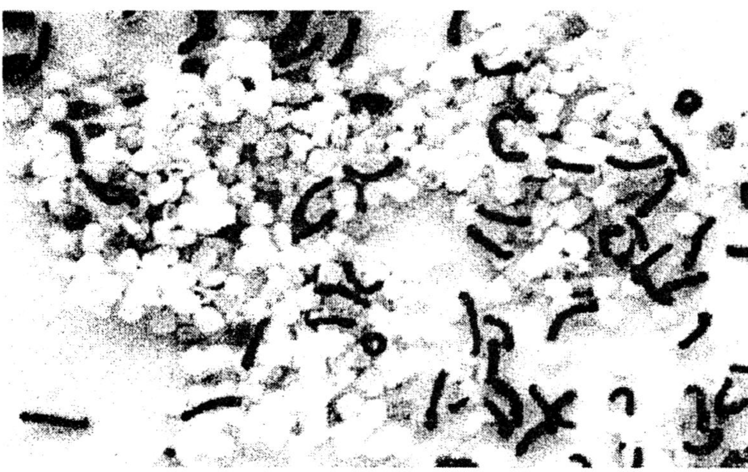

Figure 2. After proper incubation of eggs we get excellent hatching.

The first three days after the beginning of the incubation, the temperature should be maintained between 15 and 20°C (this can be ensured by the temperature of environment). During the remainder days, the temperature of the incubation room should oscillate around 25°C. In practice it has been proved that a continuous exposure of eggs at a temperature of 25°C shows good results. The heating of the incubation room can be done by a heater connected to a thermostat.

Humidity

Relative humidity has also some effect on egg hatching. Physiologically, too dry or too wet air is harmful to the eggs. For example, under dry condition, the eggs may lose their water, and this will result in erratic hatching and increase the percentage of dead eggs. On the other hand, high humidity (more than 90%) produces larger larvae which are in fact very weak.

The whole incubation period requires high relative humidity (75-80 %). The humidity can be created by a humidifier with humidity sensors or by regular watering the floor of the incubation room.

Light

As in the case of temperature light also has some influence in the change of voltinism. While incubation in darkness produces non-diapause eggs, incubation in brighter light produces diapause eggs. In some cases, temperature and light have cumulative effect on incubation and hatching of eggs. In order to ensure a uniform hatching, eggs should be exposed to light until the appearance of the first larvae. According to Pang-Chuan et al. (1988), keeping eggs in darkness can inhibited them from hatching for up to only nine (9) hours.

The light/dark ratio should be: 18 hours light/6 hours dark. When 50% of embryos have completed their development, continuous darkness must be maintained until the emergence of larvae.

Material Installations

Basic conditions of silkworm rearing are appropriate room and enough mulberry leaves of good quality. It is possible to perform 3 rearings per year (one for spring, another at the beginning of summer and the last one early autumn) depending on the region, the prevailing meteorological conditions and the way of mulberry field management.

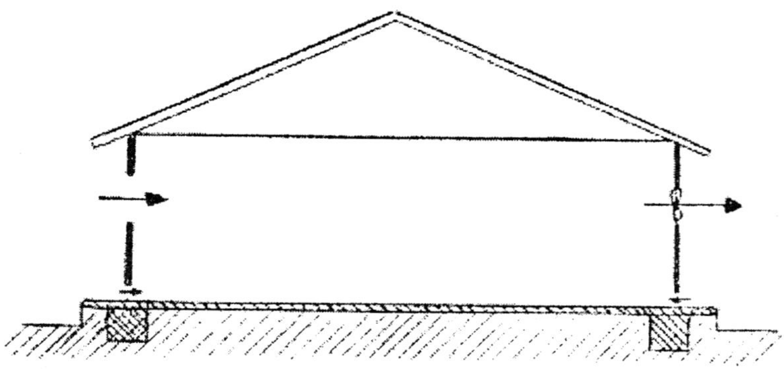

Figure 3. Silkworm rearing unit in cross section.

However, in order to achieve a successful rearing, except for diet and necessary space, suitable temperature, relative humidity as well as hygienic conditions should be ensured.

Rearing Rooms

Two types of silkworm rearing units are distinguished: permanent and temporary. Permanent rearing units are specifically constructed buildings equipped with appropriate installations and infrastructures to ensure suitable environment. Such units must also provide facilities for the execution of the necessary rearing activities. For the rearings, it can be used any space that has been constructed for other purposes (rooms of residence, burns, stables etc). This space must provide basic silkworm rearing conditions and it must be free at the time of silkworm rearing. Also, since good aeration is essential, the windows of the rearing rooms must have hardware cloth in order to prevent entrance of insects, mice, etc in the rearing rooms (Fig.3).

For surface saving, the silkworms are reared on or beds that are placed one above the other and supported by wooden supports. There are three levels of beds separated by a distance of 75 cm.

Needs in Surface

Depending on the larval age, for a box of 20,000 eggs, the requirements in surface for silkworm rearing are represented in table 1. For a normal and successful rearing, comfortable room for the rearing and growth of silkworm is a prerequisite.

Table 1. Surface requirements for silkworm rearing

Instar	Required area (m^2)
1st	1
2nd	2
3rd	5
4th	10
5th	20

Needs in Space

For a rearing unit of silkworms obtained from 20.000 eggs, the requirements in space for silkworm rearing are represented in table 2. From the above table, it appears that for the first instars, the needs in space are small. The 5th instar require more space (up to 50 m_3) but this is for few days only (8-10 days).

Table 2. Space requirements for silkworm rearing

Instar	Required space (m_3)
1st	2
2nd	2
3rd	10
4th	20
5th	50

Preparation and Duration of Rearing

Disinfection of Rearing Rooms

Rearing units, room, tools and implements are liable to infection by pathogens and these pathogens can cause serious diseases to silkworms. It is therefore necessary to disinfect the rearing room and the rearing tools. This should be done few days before the beginning of silkworm rearing. It is also recommended to carry out disinfection after completing the rearing and after next brushing. Disinfection of rearing room and units is a requirement for a successful silkworm rearing. In case of the presence of infected or dead larvae, disinfection should be carried out more thoroughly. The two simplest disinfection practices are burning and burying. Samples of infected eggs, larvae, nymphs or adults and infected materials, such as paper and bedding, should be burned and buried in the soil.

Chemicals are applied either by spraying on the surface or by fumigation. Surfaces of rearing units are sprayed while rearing rooms are fumigated. Before fumigation, the rearing rooms should be hermetically sealed. Volatile substances or substances in form of gas are used for fumigation. In practice, one of the following three disinfection methods is used:

Solution of formalin 2%

The commercially available formalin has a content of 35%

(w/v) of formaldehyde. In order to prepare 10 litres of a 2% solution, 600 ml of formalin are added to 9, 4 litres of water. Sprayed on the surface of the room, the solution acts not only as a gas but also as a contact disinfectant. For the disinfection of a surface of 10 m^2, one litre of solution is required. The tools are dipped in the solution for 10 minutes and then rinsed with water.

Aqueous solution of bleach

The commercial bleach has a content of 5% (w/v) of sodium hypochloride (NaClO). It is usually used for disinfection as a 1.5% solution. In order to prepare 10 litres of such solution, we add 3 litres of commercial bleach in 7 litres of water. For the disinfection of a 10 m_2 surface, one litre of this solution is required. The surfaces are sprayed while the tools are dipped in the solution for 30 minutes and then rinsed with water. The surfaces and the equipment should be well wetted. The solution should be prepared just before the application in order to ensure a successful disinfection. The active chloride is evaporated easily and this is the reason for the decreased effectiveness of the spraying solution. The spraying solution is effective against viruses such the Polyhedrosis virus, fungi (muscardines) bacteria, and protozoa (spores of Pebrine).

Fumigation (Potassium Permanganate and Formalin)

The room is closed hermetically and containers made of clay or glasses are used in which the potassium permanganate ($KMnO_4$) is placed initially and then formalin is poured but never the opposite. For a successful disinfection, 20gr of potassium permanganate and 30ml of commercial formalin per m_3 are required. However, we should previously moisten the surfaces with a spraying device. The mixture of these two substances releases enough fumes that should circulate in the hermetically closed room for 2 days. Afterwards, all the windows should be opened for 5 days before the silkworm rearing begins.

The disinfection is effective when the temperature is higher than 20°C. The higher the temperature and the humidity are, the better the disinfection is. In any case, the spaces should previously be thoroughly cleaned so that the disinfection can be effective. Even though the formalin is the most common and good disinfectant, when it is used to disinfect equipment and closed rooms it causes eye irritation to humans and animals. Consequently, aqueous solution of bleach is used as an alternative disinfectant of rearing rooms. However, this solution has the disadvantage of eroding the metal surfaces used for silkworm rearing. It is more suitable for disinfecting wooden and plastic materials.

It is also important that the following are stressed:

— The entry into the silkworm rearing room should be allowed only to workers of the rearing units.

— Those entering the silkworm rearing rooms should thoroughly clean their shoes on a mat wetted with a disinfectant.

— During the silkworm rearing, infected and dead larvae should be removed and burned.

— Rearing beds on which a large number of dead larvae were found should be replaced by disinfected ones and a number of dead larvae should be sent for laboratory examination.

— In order to prevent the entrance of pathogens, the floor, the corridors, the entrance and the storage room of mulberry leaves should be disinfected by spraying with an aqueous solution of bleach 1,5 % every 3-4 days.

Rearing Duration

The silkworm larva undergoes 4 moulting that determines the 5 instars of the larva. From the rearing of polyhybrid races, the usual duration of each instar and the duration of their corresponding sleep are represented in table 3.

Table 3. Duration of silkworm development

Instar	Duration (days)	
1st instar	3	(Hatching until first sleeping)
1st sleeping	1	
2nd instar	2,5	(end of 1st sleeping until beginning of 2nd sleeping)
2nd sleeping	1	
3rd instar	3	(end of 2nd sleeping until beginning of 3rd sleeping)
3rd sleeping	1	
4th instar	5	(end of 3rd sleeping until beginning of 4th sleeping)
4th sleeping	1,5	
5th instar	8	(end of 4th sleeping until beginning of cocoon spinning)
Total	26	

During moulting, the larva stops feeding, hooks on a surface with pseudopods, raises its head and remains thus motionless, until the process of moulting finishes. This stage of immobility, which is very characteristic, is named sleep and lasts 1-1,5 days. Thus in regular rearing conditions, the larval stage of polyhybrid silkworms lasts from 25 until 28 days. It should be pointed out that the larval stage is greatly influenced by the temperature and other environmental conditions.

Cares of Rearing

Temperature

Rearing temperature directly affects the physiology of the silkworms, in nutrient absorption, digestion, blood circulation, and respiration. The first three instars of silkworms require high temperature (25°C). However in the next two instars, the need in temperature decreases (23-24°C).

Humidity

Humidity influences the growth and physiology of silkworms in two ways: First, it has effect on water content and evaporation in the body of the larvae (through feeding). This parameter plays a great role in the regulation of the body temperature. Secondly, humidity has effect on the preservation of mulberry leaves and therefore indirectly on the larva's appetite. The optimum relative humidity for the first 3 instars must be 80% while in the last two instars 75%.

Air and Ventilation Conditions

Air is essential because the silkworms (particularly the 4th and 5th instars) have high requirements in oxygen. However, it is recommended to avoid air current or stream. A bad airing creates conditions that encourage the growth of fungus in bedding and pathogenic agents of the silkworms.

Light

It is recommended a moderated intensity of lighting with duration of 15-16 hours of light and 8-9 hours of darkness. Intense light or the direct solar radiation must be avoided in the rearing rooms. In practice, the duration of lighting is that of day, without the use of additional lighting.

Feeding with Mulberry Leaves

To rear a box of 20.000 eggs, we need nearly 500 Kg of leaves. The requirements in mulberry leaves for each instar are distributed as follows:

1st instar	2 Kg
2nd instar	8 Kg
3rd instar	20 Kg
4th instar	80 Kg
5th instar	390 Kg
Total	500 Kg

From the above data, it appears that the need for mulberry leaves increases as the silkworm grows. Thus, 70% of the total quantity of the mulberry leaves is utilised at the last instar. It is during the 5th instar that the most labour is required. The silkworms feed continuously on mulberry leaves. They stop feeding only when they are sleeping. That is why, in order not to delay their growth, it is recommended to provide continuously with fresh leaves. The first leaves (up to 3rd age) are supplied chopped in thin films while during the 4th instar, entire leaves are supplied. In the case of the 5th instar even entire shoots can be supplied.

The main objective of chopping the leaves is to facilitate the feeding of the silkworms. The sizes of chopped leaves vary with the larval instars, with the number of feedings as well as with the seasonal climatic conditions. Also, the more feedings a day, the smaller the leaves are chopped (as fewer feedings are required, the leaf fragments should be larger). Leaves can be chopped in square shape, rectangular shape and larger fragments.

In general, the target size of the chopped leaf is: 0.5-1.0 cm_2 for the first, 1.5-2.5 cm^2 for the second, and 3-4 cm^2 for the third instar. It is also recommended to use 1 cm wide pieces, 2 cm wide pieces, and 4 cm to 5 cm wide pieces for the first, second, and the third age larvae, respectively. Because manual leaf chopping can be time consuming, leaf chopping machines are used. With such machines, one can obtain different sizes for different instars by simply adjusting the machine.

If the right quantity of mulberry leaves is supplied and if suitable conditions of temperature and humidity are created, three feedings per day are sufficient. By covering the beddings with plastic (polyethylene), one can avoid the drying of leaves, mainly during the first instars when leaves are chopped. For the leaves of 4th and 5th age, the plastic cover is not essential because entire leaves are supplied and they don't dry quickly.

Change of Beds

Each time when the silkworms are moved off, the beds must be cleaned. The litters on whatnots (remains of leaves, impurities of larvae, etc.) must be removed. Otherwise, the litter can go mouldy, creating serious unsanitary conditions. The change of the beddings is carried out at the beginning of each instar. First, a perforated piece of paper or a net is placed on the bedding and then mulberry leaves are placed. The larvae mount up to the mulberry leaves. And then, the bedding under the paper is removed.

Thinning of Silkworms

Thinning is one of the basic cares of silkworms rearing. With thinning, silkworms grow more comfortably, are well-developed and diseases are prevented. As a guide of thinning, the table showng the needs in surface can be used.

Timing of Rearing

On a given bed, all the silkworms should be of same age. They must begin sleeping and feeding all together. If this timing is not respected, certain will be asleep while the other will need to feed. Those looking for food will disturb those which are asleep. This may cut the silk filaments with which silkworms tie their moult with the leaf and thus cannot get out from the old skin. Finally they are covered with new leaves and many of them die from suffocation. To timing the silkworm rearing, the following actions are recommended:

— The larvae hatched on the same day are placed in separate beddings and should not be mixed with the larva hatched on different day. The same should be applied when we pick up the silkworms from their sleep. In this sense, we should not put in the same bedding the first and second day larva but we should rear them in separate beds.

— In order to maintain this balance the temperature in the entire bed should be the same. If the temperature in the same bed is not uniform, silkworms in the warmer part will eat more and grow faster than those found in the colder part.

— When we observe the emergence of small silkworms from the eggs, we must wait till the next morning and then we stretch out net above the eggs. Once the small silkworms get on the tulle, we supply them tender and soft mulberry leaves. When the tulle is filled, the larvae are taken off and placed on another bed. We repeat the same process up to the midday placing all larvae on the same bed. Afterwards we supply them (all together) their first meal. The silkworms hatched on the next day are placed in separate bed before providing the meal. On the third day, we proceed in the same way. In case of the presence of eggs which are not hatched yet, we create fourth bedding. The silkworms come out only in the morning; that is why, we do not collect them in the afternoon. If the incubation takes place normally and the temperature remains at 25°C, then the incubation should be completed at a maximum period of 2-3 days.

— For each meal, mulberry leaves should be uniformly spread on the bed. Because mulberry leaves given young larva are supplied chopped, they dry quickly. That is why, feeding with these leaves should be more frequent and in small quantities. At the older ages when entire mulberry leaves are supplied, feedings are less frequent but the quantity of leaves is larger than of the previous feedings.

— When the first larvae are starting to sleep, we decrease the quantity of leaves. With the completion of sleeping of the first silkworms, we stop feeding. When half of the larvae complete their sleep, we line up leaves with their twigs and the larvae get on leaves. Then, we move them to a new bed. Those that are found to be still asleep are left for 24 hours without feeding them and when they wake up we create for them a separate bed. The separation woken up larvae from those that are still sleeping is accomplished with the perforated paper or net mentioned above. With this process the thinning and an easy change of bedding are simultaneously accomplished.

— In the rearing room, we usually have three levels of beds separated one from the other with a distance of 75 cm. The first woken up larvae are placed in the lower level on the first day; the second on the intermediate level and the third day on the third level. If mulberry leaves are supplied at the same time, we will observe that the silkworms on all three beds will sleep simultaneously. This happens because the bed at the upper level is warmer than the one on the lower level. Larvae on the upper level eat more and consequently grow faster.

— The presence in rearing beds larvae of different size in a silkworm rearing unit is recommended. This can help to overcome the problem of increased labour in sericulture mostly during the process of mounting. However as mentioned above, it is important that silkworms on the same bed are of the same age.

Process of Mounting

When the silkworm larvae complete their development, they become uneasy, stop feeding and begin releasing large drops of gastric liquid. Their body becomes smaller and transparent. These processes are evidence that the larvae have completed their feeding and will proceed to the pupal stage, thus beginning the phase of mounting (cocoon spinning). At this stage, we must place on the rearing beds suitable materials used as surfaces for cocoon spinning (Fig. 4). As mounting materials can be used dry twigs of heather, thyme or of other bushes (without thorns), artificial twigs etc. (Fig. 5 and 6).

The process of mounting is also influenced by environmental conditions such as temperature, humidity, light and air. Humidity influences the quality and quantity of the produced cocoon as well as its reelability. During mounting, the temperature in the rearing room should oscillate between 23-25 °C and the relative humidity must be around 70-75%.

Figure 4. Cocoon spinning

Mature larvae are negative phototactic and tend to turn away from light. When a strong light enters the mounting room, the larvae will crowd to one side, forming many double cocoons and cocoons of uneven thickness. It is therefore recommended to have moderate and uniform light in the mounting room.

Cocoon Harvesting

Silkworms need 2-3 days for cocoon spinning and 3 more days during which larvae change into pupae. Thus, 8 -10 days after mounting, picking or collection of cocoons must begin. The weak cocoons should be separated. When picking is completed, the exterior fat spinners from the good cocoons are removed. These spinners are becoming strong thread.

Figure 5. Mounting on a branch (left) and on plastic net (right)

Cocoon Cooking

After a maximum period of 15 days from the mounting, cooking of the cocoons (that is to say the killing of pupae) must take place. Otherwise the pupae will become adults (butterflies) which make holes on the cocoons to come out. These cocoons, due to the interruption of the continuity of the fibre can not be reeled and thus are useless.

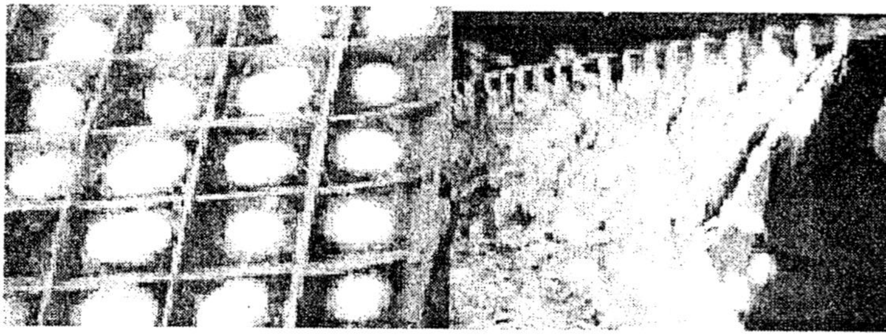

Figure 6. Square frame (left) or peaces of string tighten on a stick (right) mountages

There are various ways of cocoon cooking. The most common ways are:

— Wet cooking with steam (Fig. 7)

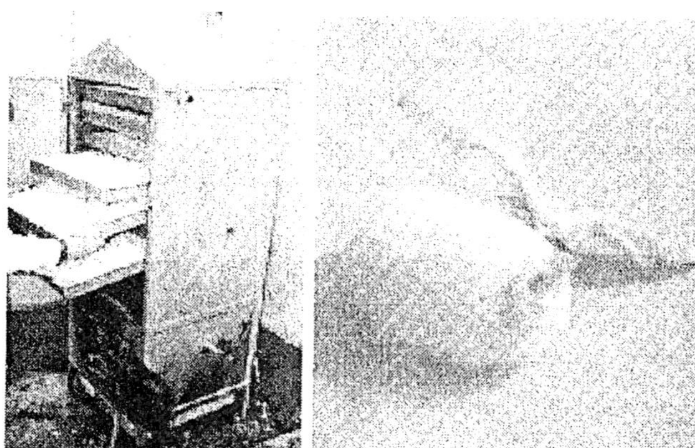

Figure 7. Cocoon cooking (left) and silkworm eclosion (right)

— Dry cooking with dry—hot air

— Cooking with phosphine

In the first case, cooking is carried out with steam that is produced in special steam boilers and that are channelled into the cooking chambers. These chambers must be hermetically closed. They should have in the roof a pipe used for steam escape and they have supports where metal drawers are inserted. In these drawers, cocoons are spread in layer of 15 – 20 cm thick.

In the second case, cooking is performed with the utilisation of dry – hot air, produced also in special ovens. In both cases the temperature should be maintained in the range of 80-90°C for 30 min. The temperature must be checked in order not to exceed this limit. After cocoon cooking they are spread for drying so that they are ready for reeling (Fig. 8).

In the third case cooking is performed with phosphine. The chamber should be hermetically closed. In this case, close attention is recommended because phosphine is a very poisonous substance. It has however the advantage of protecting the cocoons from the moth (Dermestes sp.) and are reeled easily with a better silk quality.

Mulberry Leaves

Mulberry leaves are the unique food of silkworms and without mulberry leaves, it is almost impossible to establish a sericultural unit. There are also other plants on which the silkworms can feed. Moreover, with some of these plants, the silkworms reached the cocoons spinning stage. However, this alternative food does not ensure a productive sericulture. Sericulture could not be developed without suitable cultivation of mulberry trees and the production of leaves of good quality and large quantity. The mulberry tree grows in tropical, subtropical and temperate climates and can stand temperatures of 25°C below the zero. Mulberry tree has the ability to grow on unfertile and less moist soils, however when it is cultivated on fertile soils with regular irrigation it produces large quantities of leaves of the most excellent quality. In order to produce large quantity of leaves, a mulberry tree should have many branches of leaves and should grow fast.

There are many varieties of mulberry trees, each of which has its particular characteristics. Despite the differences between the varieties, there are some standard attributes that should be respected in order to be recommended for sericulture. These attributes are: resistance to diseases and pests, resistance to drought and adaptability to a variety of soils. Varieties with high sericultural value which are cultivated in Greece are:

Proussis, Kairyo Ichinose, Kokerka, Heyebai, Kokuso 20, Kokuso 21, Kokuso 27, Fengchisan. There are also many local varieties.

Mulberry Propagation

Mulberry is propagated innately (seed) and rudely. There are many ways of rude propagation such as cuttings, layering, grafting and tissue culture. The cutting is the easiest way of mulberry propagation.

Mulberry Plantation

In order to ensure a satisfactory leaf production, a modern mulberry plantation should have at least 1600 trees per hectare (planting distance of 3X2 metres). The dense planting has the following advantages:

- Complete mechanisation of the mulberry cultivation, as digging, fertilization, irrigation etc.
- The dense foliage does not allow the growth of weeds on the ground; therefore after June and until October weeding will not be needed.

— The mulberry is considered the ideal tree for fire protecting zones.

— At a recommended density the largest leaf area (photosynthetic) per unit of surface is produced.

— Leaf production is maximised: on the average 10-15 kg per tree (after the 4th year) or 20,000-30,000 kg per hectare.

— If few trees are planted (e.g. 28-40), it is not possible the uniform coverage of the surface. This may result in the growth of weeds between the trees. Dense planting provides better protection from soil erosion.

Planting pattern in temperate climate is dictated by mulberry varieties, types of soil and cultivation practices etc. According to Lim et al. there are three basic planting patterns (Fig. 1):

— single rows in high plant density, 1.5-1.8X0.5-0.8m (about 9,000 trees/hectare).

— rows with paired plants, wide ride 1.8 m, narrow ridge 0.6 m X 0.5 m (about 16,000 trees/hectare).

— single rows in lower density, 2.5X1.5 m (about 2,500 trees/hectare)

Mulberry Cultivation

As in the cases of other plants, in order to be productive, mulberry trees must grow under optimum conditions.

Soil

Mulberry trees grow on argillaceous fertile, well-drained soils with good pH 6.5 – 7.5. Addition of organic matter (compost and others), enriches the soil with nutrients, improves its structure and increases the action of micro-organisms.

Figure 8. Modern cocoon reeling machine

Irrigation

Despite the fact that mulberry trees resist to drought, their irrigation increase leaf production. For summer and fall rearing, irrigation is very important since we ensure optimum leaf moisture, of course with the combination of the suitable pruning.

Fertilization

Fertilization increases the quantity and improves the quality of leaves. The type and the quantity of fertilizer depends very much on the soil fertility and on the irrigation. When we have many rearings per year, fertilization should be frequent. Usually, for a medium fertile soil, the average quantity of fertilizers is 230 kg of

N, 150 kg of P and 170 kg of K per hectare, applied in three doses.

Tillage- Weeding

The growth of weeds is influenced by the density of planting, the irrigation method and frequency etc. To destroy weeds, we apply physical as well as chemical methods (herbicides). While applying pesticides, particular measures must be taken not to drift to mulberry leaves. Pesticides on mulberry leaves may have negative effects on the reared silkworms.

Pruning

Depending on the height of trees and the leaf collection methods, various types of pruning are applied. From the planting of the tree up to the 4th-year tree, only shaping pruning is performed. After the 4th year and according to the number of rearing units harvesting pruning is applied. Depending on the density of planting, three types of cutting are distinguished: low cutting (plant cut at a level less than 0.5 m from the ground), medium cutting (plant cut at a level between 0.5 and 1.5 m) and high cutting (plant cut at a level more than 1.5 m).

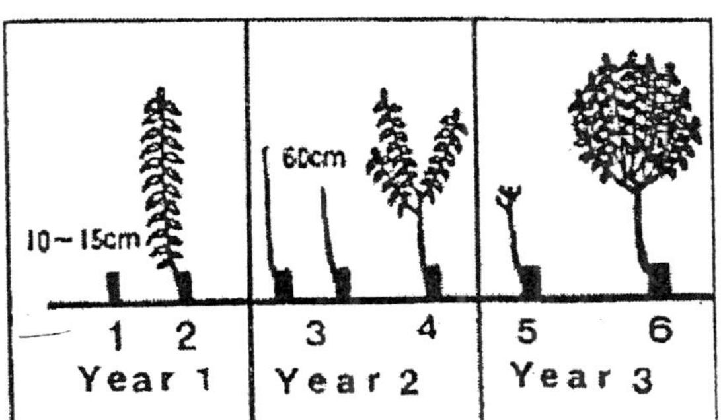

Contd....

Contd....

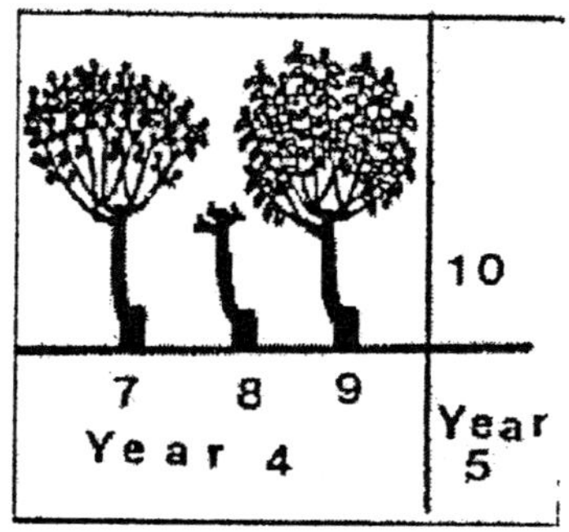

Figure 9. Medium cutting pruning, 1. Planting, 2. Late fall, 3. Spring cutting, 4. Late fall, 5. Spring cutting, 6. Late fall, 7. Spring, 8. Summer cutting, 9. Late fall, 10.Same as in the 4th year

For example for a mulberry plantation of 2.000 trees per hectare (2.5 x 2 m), the most suitable type of cutting is the medium cutting as explained in figure 9.

3

Growth of Silk Production

Long Term Growth of Silk Production

Long term growth of aggregate demand for silk textiles in Western Europe and elsewhere where only partially satisfied by growing imports from South and East Asia. It had a clearly discernible effect on Mediterranean Europe and Near East countries silk production too. Large growth of silk production took place in Southern Italy during 15th to 17th Century (Calabria, Sicily) followed by a real boom in silk production in Northern Italy (16th to 19th Century). Italy thus became the largest silk producer among Mediterranean countries from late 16th Century onwards, covering some 2/3rds of their aggregate silk thread output.

By middle 19th Century Italy was the world second exporter of silk after China. Beside Italy, relevant increases took also place in Valencia and Murcia areas in Spain (15th to 17th Century), on the mountain slopes of coastal Syria and Lebanon (16th to 19th Century), in the Ottoman Empire (Bursa, Edirne, Thessaloniki) and as far as present day Azerbaigian (Nukha) and the Caspian shores of Persia (Ghilan). A large portion of exported silk thread output from such areas was directed to Western

Europe (later on, to USA) silk textile producing centers. Oddly, most of these centers were situated in areas where little or no silk cultivation existed, such as Lyon, Tours, Paris, London, Macclesfield, Vienna, Zurich, Amsterdam, not to mention silk textile factories which were developed in Sweden or in Moscow.

Characters of Silk Production

Basic Inputs

Among the elements that characterise silk production one should mention, firstly, its rigid connection with (optimal) mulberry trees growing areas. As a matter of fact. silkworms can be fed with fresh mulberry tree leaves only. It is therefore imperative that silkworm raising be conducted in areas where an easy-to-reach, abundant supply of mulberry tree leaves be available. Moreover, mulberry trees can withstand repeated leaves spoliations under a rather favourable climate only: early cold spells may strongly damage leafless trees.

Therefore the existence of mulberry trees does not imply per se the practicability of a profitable silk cultivation (such as in Northern France or in England where mulberry trees may be planted with ease, but cannot be regularly stripped of their leaves). (to be further elaborated) Since the 15th Century, several attempts were repeatedly made to acclimatise mulberry trees varieties whose characteristics could maximise leaves output, resist sudden climate vagaries, ease leaves stripping operations etc.

Successful man-guided and man-made interventions on mulberry trees were often instrumental in deeply altering the pace of development of silk production in certain areas, such as with the introduction of Morus

Alba in Western Mediterranean in early 15th Century or with the cross-breeding of Chinese mulberry tree Lu in Japan in late 19th Century allowing for multiple leaves cropping.

Similar considerations may be made in regard to silkworm varieties, their cross-breeding and their adaptation to local climates, and to silkworm eggs and their optimisation of choice and sorting, preservation methods, long distance transport techniques etc. "Rejuvenation" every few years of local silkworm races through long-distance import of highly reputed silkworm eggs was a must in most advanced silk producing areas around the Mediterranean Sea up to early 18th Century and even later. Calabria and Valencia stocks were seen as the best ones, Lebanon too gained much public favour.

When the pebrine silkworm epidemics struck European silk cultivation in the middle of 19th Century, ravaging it for some 25 years, Italian and French traders roamed, with much disappointment, the whole world unable to find a place from where to import, year after year, huge quantities of good healthy silkworm eggs. They eventually found it in Japan only, giving a big boost for some 15 years to its terribly weak balance of trade and at the same time saving an occupation which in Italy gave bread to millions.

Market Value

Silk (silk thread) enjoyed a very high unitary value (selling price/weight) throughout the period, a gradual decline taking place from around 1870s onwards only. It was by far the richest product a rural family could produce on its own, and such a consideration extends back to (good) cocoons. In Western Europe cocoons and, to a certain extent home-made silk thread hanks (raw silk) too, sold for prompt cash. It took less than two months (May to early June) to get a cocoon crop and a couple of weeks more to reel raw silk hanks out of them. Thus cash flowed in into peasants hands in the most precarious period of the (Western European) agricultural year, when Winter food reserves had been worn out and Summer grain crop had not yet been harvested.

Elders of a North Italian town commented with much satisfaction in early 17th Century on how the recent spread of silk cultivation in their countryside had made disappear the hordes of starving peasant beggars that used to roam their town streets in early Summer. King Henry IV spent huge sums to import from Spain and plant in and around Paris and in other outlets of Northern France tens of thousands of mulberry trees in order to free from misery his peasants and get rid France of the need to import costly silk from Italy and Spain.

Geographically Unlimited Market

As a consequence of a very high unitary value of basic silk products (cocoons and silk thread), very little if any of them was kept home and most of it was brought to the market as early as in Medieval times. Merchants combed every corner of the country to collect silk thread to be used in far-away urban textile centers or to be exported to distant lands. In fact they bought even minute amounts of waste silk, highly appreciated for less expensive or mixed silk cloth. A Zurich firm involved in mixed-silk cloth production for North European customers used to send its agents as early as in late 17th Century into scores of French and Italian Alpine valleys - from Friuli to the East to Ardéche to the West - to collect in each place tiny amounts of waste silk to be reprocessed in Zurich Thus silk production linked silk producing rural units to distant markets and to cash economy even in remote and secluded places where everything else was heavily linked to self consumption and to sleepy local market circuits.

High quality raw silk fetched much higher prices and was always in high demand. Merchants were ready to go anywhere to get it. During the century-long Pax Mongolica traders had the exceptional opportunity to reach China without hindrances. An early 14th Century merchants' manual of one large Florentine trading and banking firm, tells us in detail how to go by land to China (with a 20-odd months return trip on horseback and foot) to buy raw-silk. It has been estimated that one third or more of silk thread that Lucca employed for its fabled luxury textiles in late 13th Century came from China.

Supranational Competition

The ease with which lightweight expensive raw silk could be moved around the globe under any circumstances, implies that nowhere in the Western world was raw silk production free from the challenge of competing raw silk of higher quality and/or of lower price. In a broader sense it appears that no guild, state or sumptuary regulation was ever able to seriously block or impede for long massive arrivals of competing products, or to suffer dire consequences by failing to adapt to the pressure of competition.

Those countries or areas that failed to come up with or who were unable or unwilling to match competing products were either kicked out of the market or cornered in much less profitable niches. Such were the cases of Grenada's silk production in late 16th Century, of Sicily and Calabria during 18th Century, of Tuscany in early 19th Century. Spain attempted in vain to insulate its American colonies from raw and woven silk smuggled in from China via Manila.

Silk Thread Making Technology

Contrary to a widely held view, raw silk making technology, although apparently rather simple, has by no means been stagnant in the long run. Indeed, under the driving forces of a harsh, ever present, external competition, efforts have been constantly made to improve it and to adapt it to variable final users' request and to improve quality and productivity in order to maximise returns.

Technology historians, fascinated by the dramatic improvements to productivity brought about by entirely new, huge, costly complex mechanical devices of the Industrial Revolution have at times failed to grasp the crucial role played in productivity and/or quality enhancing played by small, cheap ingenious devices aptly inserted in a long time experimented low-technology chain of production. Contemporaries were indeed very well aware of the latter's role.

Some of the foremost French technicians fought for decades during 18th Century in order to devise a reeling machine able to match the Piedmontese one - a mostly wooden simple structure which could be built by a good carpenter in a few days work. Crossing of a silk thread on itself on its way from basin to reel is a simple hand operation requiring a few seconds to be completed, yet there are scores of brevets and patents in Italy, France and England in the first half of 19th Century on how to best perform an operation which every expert agreed could enhance the market value of silk thread, if correctly performed, by up to 50%. An absolute must in high quality silk thread was its perfectly constant diameter. Hand testing of the diameter of a silk thread of 0.1 to o.2 mm and the act necessary to keep it constant, is an incredibly difficult task which required exceptional dexterity on the part of the master throughout the working day. In fact mechanisation of thread testing has been one of the most challenging task to mechanical engineers up to the early decades of 20th Century. It was only then that the fingers sensitivity of a master's expert hand could be safely substituted by automated mechanical devices in silk-reeling plants.

Skills

Silk-reeling, being the chain of low technology steps which leads from cocoons to silk thread, was mostly performed in rural houses or in small, often home-made, plants usually employing a score or two of workers up to late 19th Century. In order to obtain a high-quality, highly appreciated silk thread, its (apparently) simple devices had to be operated by a highly qualified, highly skilled manpower. Dexterity, attention, care, and above all, solid experience dearly bought in many years of harsh apprenticeship, were essential elements for success.

In Italy, to be recognised as a master reeler required, as early as in 15th Century, from 4 to 7 years of apprenticeship and to pass a difficult entrance test in front of a commission made up by expert traders in silk. Such rules were not a guild barrier to limit access to trade, but an inescapable step to insure much needed stable quality (as defined by several sophisticated parameters) to silk threads employed in luxury cloth making.

Piedmontese rules of 1670s detailed each and every step of silk-reeling process, both in regard to every minute part of machinery and connected devices as well as to masters' qualifications and duties, under harsh penalties and strictly enforced state control. At the same time it imposed to silk reeling plant owners not to have masters work on piece-rate system, but on daily wages only, thus recognising that haste and quantity were lethal enemies to quality.

As a result, Piedmontese raw silk enjoyed a virtual monopoly on European markets in its highly appreciated organsine silk thread for some 150 years. All cut-corners attempts made during 18th Century in Southern France, around Valencia in Spain, in British North America or in Bengal, trying to lower skill and/or machinery ingeniousness in order to obtain - cheaply and soon - high quality raw silk equal to the Piedmontese one, failed miserably.

Skilled Manpower Migration

The quest for improvement in silk thread making led to innumerable cases of individual movements of highly qualified personnel to other places. Inversely, it lead an equally numerous number of instances in which individuals slipped in into foreign places in order to acquire guarded skills or "secrets" to bring back home and/or to resell to interested parties. Examples of the former go, just to mention a few of them related to Piedmont "supremacy", from one Piedmontese technician of late 18th Century employed in Naples and subsequently in Calabria to instruct local people in Piedmontese reeling techniques, to the score or so of Piemontese experts smuggled out together with machinery and drawings around 1760 by East India Company agents to be sent to Bengal in order to improve local silk making, to Piedmontese entrepreneurs lured to North Portugal by Lisbon Government officers in the 1770s, to a Piedmontese family of silk experts invited in the British colonies of North America, to an individual Piedmontese adventurer building a silk-throwing plant after Piedmontese design for Mehmet Alì in Egypt in the 1830s.

Gernder Specialization

In talking of silk reeling masters one should mention that they were all female masters. Indeed, most operations of the first section of silk production cycle were traditionally performed by women only, notably the most delicate of them - silkworm raising. From China and East Asia westward, through Iran, Caucasus, Lebanon, to Mediterranean countries, the role of women in silk production has been paramount in the whole period under review. Female knowledge of how best to handle extremely delicate insects (silkworms) or equally delicate passages in silk reeling have been often stressed or plainly acknowledged by contemporaries.

In examining the long string of didactic poems and manuals devoted to silkworm raising published in Western Europe from late 15th Century to early 19th Century, one is struck by the fact that the absolute majority of them is dedicated to a woman and that many of their authors (invariably men) either recognise their debt of knowledge to a woman or beg pardon to women for describing a trade they, women, know best.

Hidden in archival documents are also practical proofs of such recognition: as it is in the case of an Italian couple hired by a Spanish employer in 18th Century to supervise silk.making operations, the weekly salary of the woman being by far higher than that of her husband. Moreover, in several instances when attempts have been made to pass into male hands operations which were traditionally performed by women, quality and productivity declined to such an extent to render it necessary to revert to women's work. Such is the case of a large silk firm in Tuscany in the early to middle decades of 19th Century. Inversely when in late 18th and early 19th Century in Sicily and in Calabria a few entrepreneurs attempted to modernise silk reeling (which - an exception in the whole Mediterranean - had been traditionally performed by migrant men masters only) they made use of women labour. Lower women salary compared to male levels did obviously count in their choice, but the progress obtained in quality performance appears to have been quite considerable.

The common expression calling the woman in charge of a silkworm breeding room in China as the mother of silkworms is intriguingly reflected in several 16th and 17th Century Italian poems on silkworm raising in which the body of the woman in charge of them is called more apt for its natural warmth and youth force to raise those insects than a man could ever be.

A number of beliefs and of so-called superstitions that have traditionally surrounded the extremely delicate process of silkworm raising, as well as practices with special herbs to fend off from silkworms sickness and nuisable enemies are strikingly similar from inner Asia to Spain, and almost always mastered by women only. It is suggested that in moving from China to the West in a several centuries long march along the Silk Road, silkworm raising techniques and related rituals might have kept the femalecentered character they originally had in China.

4

Use of Bamboo in Sericulture

Bamboos form the backbone of agriculture through agricultural implements, tool handles, fencing, transport, storage and processing equipment of all agricultural produce in tropical India. Similarly, bamboo has its multifarious uses from baskets for collection of mulberry leaves, harvesting, storage, rearing of silkworms, production, transport, storage and processing of cocoons, filatures, reeling and weaving of silk. The different uses of bamboo in sericulture are described below:

Leaf Collection Basket: The basket is specially designed for collection of plucked mulberry leaves. It has a small mouth on top and a wide middle, sloping again towards the bottom to take the shape of a compressed sphere. It is about one kg in weight and can hold 12 kg of leaves for easy carriage. It is woven from a combination of 1 to 2 cm wide slivers between 0.4 cm thick strips and strengthened at the mouth by similar strips. It is valued at Rs.40/- each. It has an average life of one year.

Rearing Trays: Circular, with diameters of 0.9 mt to 1.3 mts and depth of 5-6 cms, it is used for rearing worms. It is priced at Rs.40/- to Rs.60/- each depending on diameter and season and it weighs 2.5 to 4 kg and is used

5-6 times each year and has average life of three years. Bamboo trays are preferred as they absorb the excretions of the worm and are easy to clean.

Mountage: Called as "Chandrike", it is used for rearing cocoons between woven 5-6 cms strips fixed at 3-5 cms apart in a circular fashion from centre to the periphery of a bamboo mat of size 1.8x1.2 mt, by means of coir rope. It gives anchorage to the silk thread as the larvae weave the nest around them for the pupal stage. The mat is strengthened on the hind side with 3 cm bamboo strips or rounds on all four edges and criss-cross by coir rope.

The mountage weighs 10 to 15 kg and costs Rs.150/- to Rs.250/- each depending on quality and season. It is available on hire during the season at Rs.5/- per day and is needed for 3-4 days at a time, till the cocoons are removed after completion of the spinning that takes around two days. It is used repeatedly and has an average life of three years. Maximum use of bamboos is on account of use of mountages. It is explained that bamboo provides the best absorbant surface for quick driage of secreted silk to become a strand.

Mats: Bamboo mats of 1.8x1.2 mtrs are used for chopping mulberry leaves meant for feeding the worms without soiling them. Mats are also used to hold basketful steamed cocoons after they are stifled on steam in woven baskets.

Stiffling Baskets: The basket with wide top and narrowing bottom is used to gather cocoons and stifle them on steam baths so as to loosen the strands for separation of the silken threads from the pupae. The basket is about 75 cms tall, 55 to 60 cms dia at the top and 35-40 cms at the bottom. Each basket weighs about one kg. Each basket can hold 12 kgs of cocoons. Each basket has a life of 18 to 24 months.

Cocoon Storage Trays: Made of woven bamboo to diameter of about 75 to 100 cms and 6-7 cms deep. Each tray weighs about 2.5 kg. Each tray can hold 2.5 kg of cocoons. Rectangular trays of size 1.2x60 cms are also in use. These trays are filled with steamed cocoons and mounted on stands vertically for purpose of airdrying of cocoons.

Sticks in Rearing Stand: Where rearing stands are used, round bamboos of about 3 mtr length are used to support the rearing trays. The stands are made of wood whereas partitions are created by round bamboo sticks, so as to hold 10 trays vertically in the stand.

Chawki Sticks: 1 mt long 3 cms thick bamboo split or small rounds cost Rs.2/- each. Five sticks are used per ha of mulberry cultivation.

Reeling Sticks: 0.5 mt long splits of bamboo are used in reeling silk thread around them for later use in twisting into yarn or in weaving. Six such sticks are needed per ha per annum for which the cost will be Rs.6/- per ha.

Chop Sticks: 20 cms long in the size of a chopstick, these are used to pick up the delicate worms for shifting or transfer as they are not handled by fingers, due to fear of infection. However, their use is too little and not of significance.

Bamboo is a renewable natural ligno-cellolosic resource with the advantage of annual production and growth. The rhizome i.e., the underground stem of bamboos spread in the soil and annually produce culms or poles from the axillary bud of the rhizome. Emergence of culms takes place in the early part of monsoon in South India, which is June to August. The food material stored by the woody grass in its rhizome, is utilised to activate the axillary bud during early rains, when the climatic and soil moisture conditions are the best for the growth.

Emerging from the soil as a conical shoot covered by whorls of leafsheath, the elongation of the shoot is rapid. The maximum height and girth of the shoot is attained within a short period of 8 to 12 weeks. The number of culms thrown out annually depends on age of the rhizome, from the time it has developed from either seed or propagule, the soil moisture and fertility conditions. Similarly, the height and diameter of a culm for a given species or provenance, depends on the age of rhizome, food stored by it during the previous year's growth and the conditions of climate and soil. It is the annual productivity of culms which can provide sustained supply of bamboos to end uses.

The culms or poles can be split easily by knives. The split bamboo can be cut into slivers of required thickness to facilitate weaving. Since the slivers can easily bend without breaking, it is possible to use them in weaving into flat mats duly strengthened by splits or into baskets of different shapes and sizes. After evaporation of excess moisture, the airdry bamboos or splits or slivers do absorb some atmospheric moisture depending on the relative humidity. But this doesnot cause any major changes in the products like mats, baskets, trays etc. either in shape or size.

Air dry bamboo splits and slivers are absorbant because of which the secretion from silk glands dries quickly and provides a good surface and support for the strands of silk. Being locally available from natural forests and from farms in South India, bamboos become the ideal material for the various needs of equipment for silkworm rearing to silk production. Bamboo is light compared to metals and is a cheap source for use in silk industry. Since the equipment in worm-rearing and silk production requires to be handled by human labour, the advantage of volume to weight ratio makes it popular in use in South India. Besides, bamboo has very good insulation property and is not much affected by heat and cold. It is a bad conductor of electricity.

Naturally grown species of bamboo mostly used in South India consist of:

a) Medri: Dendrocalamus strictus which is common in the deciduous forest zone which is dominant in the state

b) Dowga: Bambusa bambos growing in moister zone of peninsular India and

c) Marihal or Shame: Pseudoxytenanthera stocksii which is common in farms and homesteads in Konkan region.

Two other species commercially available are:

i) Reed bamboo - Ochlandra travancorica in Kerala state and

ii) Reed bamboo - Ochlandra talbotti available in the moist pockets in western ghats.

These two species are also used in weaving, but have little application in sericulture industry. In addition to these local species, some mats are brought from North-East India for use in Chandrike. These mats are woven from Dendrocalamus hamiltonii and Melocanna baccifera.

Medri forms the bulk species in use in silk industry, followed by Dowga and Marihal. Medri is used in the round, split and sliver forms for various applications in the industry. Due to bulk availability and presence of traditional weavers in the wet zones, Dowga is mainly used in weaving mats and baskets, much less as splits and not in the round form, except in frames constructed to support rows of trays stacked vertically. Shortages of the above two species are made up by using Marihal, both in weaving and as sticks.

Annually, some bamboo mats are brought from N.E. India, mainly from Assam. These mats come in size of 2 mt x 1.2 mtrs suited to construction of Chandrikes. These mats are superior to the locally woven mats in quality, both due to the species of bamboo used and superior skill of weavers there. An average of 100 M.tonne of such mats are transported by lorries over a distance of nearly 2500 kms for use in Karnataka. Each lorry brings 3500 to 4000 mats which are sent in bundles of 35 each in the size of 2 m x 1.2 mt.

Karnataka has higher area under D.strictus than B. bambos. However, the volume or weight of the growing stock can be equal between the two species, as each culm of B.bambos is nearly three times in weight. In Tamilnadu and Kerala the proportion of B.bambos is higher. But in Kerala, reed bamboos are not much used in the equipment for sericulture. In Andhra Pradesh, D. strictus is over 90% of the growing stock. Due to higher availability in Karnataka and Andhra Pradesh. D.strictus has higher use in sericulture and is estimated at 70% by volume.

Management

Bamboos are extracted on selection basis in a felling cycle of 3 or 4 years. Bamboo forests are divided into felling series and each felling series is divided into 3 or 4 equiproductive coupes for felling. Generally, all culms which are 3 year old and above and a part of second year old culms are extracted during the year of felling. The clumps get rest for 2-3 years for recuperation. But the method of extraction has been unscientific, due to which there has been deterioration in extent and quality of growing stock of bamboos.

Upto 1990, the removal of bamboos was given on contract by the department. Pulp mills engaged their own extraction contractors on tender basis. Some cooperative societies of Buruds were engaging their own felling contractors. Since the extraction rates were decided on lowest tender basis either on stacked volume i.e. cum or on rates per 100 bamboos extracted, the bidders were looking to extraction from less congested clumps and easily accessible areas. This system resulted in unscientific removals, congestion of clumps and affected recruitment and regeneration.

Hence, the forest department switched over to departmental extraction. But the results of such working in the bamboo forests of all the states donot show any improvement in the growing stock. Grazing and fire are common in bamboo forests. Both these factors damage regeneration, affect recruitment and cause congestion of the standing clumps. No serious effort has been made by the department to either arrange for scientific extraction or proper care of the clumps to improve their condition, which is the key to higher recruitment and better shape and size of culms available for extraction. The State Governments issue orders annually or biennially for fixation of rates for supply of bamboos both to the industries and the artisans. Rates for supply to artisans involve some concession.

It is over ten years since the forest department has done away with contract system of extraction of forest produce. Forest Development Corporations have been established in all major states and they are also taking up extraction of bamboos. Yet, neither the forest department nor the forest development corporation has established scientific bamboo management, preservative treatment or organised rational utilisation which is necessary for conservation of this important resource.

The artisans and users including the pulp industries are constantly suffering due to the disorganised supply and continuous shortages of both quality and quantity for proper enduses. It is an irony that even studies are not initiated to quantify the needs of various end users of bamboo or sources from where such supplies are coming. This is the main reason because of which, accurate data on utilisation of bamboos in silk industry is not available. Even the department of sericulture has not made efforts for organised supply of bamboos their products to the sericulture farmers. Substitutes to bamboo equipment have been attempted with no tangible success. Substitutes are neither practical, bio-degradable nor energy efficient.

Storage

Immediately after extraction, the bamboos are stacked in the forest on motorable roadsides for a few days, before inspection, accounting and transport is arranged. This system applies to bulk removals either to designated forest depots or to the end users like pulpmills and artisans' cooperatives. Local villagers and artisans manage to get licences for head load or cart load removal by getting prepaid permits either from the forest department or by the village forest committees. In all the states, the artisans have formed cooperative societies. These societies are allotted bamboos at concession rates at quotas fixed per family.

Generally, 50 bamboos per family are allotted per month. The rates for supply are different to the cooperatives, market users and industries. Industries either purchase bamboos stacked in the depots or are permitted removal of dead bamboos on payment of a rate fixed per tonne. The tonnage is decided either on weight recorded at the weighbridge or average volume (cum) fixed for metric tonne. Sometimes, bamboo extracted departmentally and stacked in depots remains without disposal for months.

No arrangement is made either by the forest department or by the forest corporation or the endusers including the pulp industries, to provide preservative treatment during storage. Fungus and insects cause damage to extracted bamboos and loss thereof is neither cared for nor remedied.

Marketing

Like timber and fuelwood, there are bamboo merchants in most cities and towns of southern states. Large cities have traditional marketing centres, called "Bamboo Bazaars". Both the bamboo merchants and the leaders of artisans cooperatives get permits from the forest department for transportation of extracted bamboo, grade them according to species, lengths and sizes before storage, and sell them in retail sale to users. The sale price at the market end is 3 to 4 times the standing value of bamboos.

As most artisans are not rich enough to buy bamboos in bulk and store them for use, they are at the mercy of bamboo depot owners and merchants for their daily or weekly requirement of bamboos for making products for sale. Though there is an effort by sericulture department to establish service centers for supply of bamboo equipment, these centers are also dependent on traditional bamboo merchants for supply.

Bamboo Resources

Forest area containing bamboos in India is 8.962 million ha, which is 11.71% of the recorded forest area. The growing stock is estimated as 80.5 million tonne. Seven states of N.E. India contain 66% of the growing stock. Madhya Pradesh accounts for 12% of the growing stock. The states in South India have less stocks. According to available information, the areas and growing stock of bamboos in southern states is as follows:

State	Extent/ Sq,kms	G.S Mill. Tonne	Plantation (ha)	
Karnataka	4,920	3.5	14,400	until 31-3-2000
Andhra Pradesh	8,625	5.0	55,000	until 31-3-2000
Tamilnadu	3,100	2.0	NA	
Kerala	517	4.0	2,606	
Goa	249	2.0	Mostly planted	
Pondicherry	Nil	–	–	

Bulk of bamboo is available in the government lands and mainly in the moist deciduous forests. Dowga is prevalent in the moist vallies of deciduous forests. Marihal bamboo is grown in private lands including homesteads in the Konkan belt from Ratnagiri district in Maharashtra, through Goa upto South Kanara district in Karnataka. These three species constitute entire resources for uses in sericulture.

In Kerala state both the major species of bamboo are grown in homesteads. According to one report made for forest resources of Kerala, the annual output of bamboos from homesteads is 10,000 tonne p.a. The average annual output of bamboos from the recorded extraction as maintained by the forest department is given below.

Karnataka	70,000 MT
Andhra Pradesh	2,70,000 MT
Tamilnadu	10,000 MT
Kerala	88,000 MT
Goa	10,000 MT

In 2002, the studies made by the author have revealed that about 15,0003 tonne of bamboo mostly of Marihal variety are brought into markets in Belgaum district by the farmers of Konkan in Goa and Maharashtra and sold annually to weavers in Karnataka and neighbouring districts of Andhra Pradesh and Tamilnadu. Apart from weaving into baskets, this bamboo is used in making equipment used in sericulture. Small quantities of medri bamboo comes in from Andhra Pradesh into southern districts of Karnataka either as poles or in the woven form for use in trays and Chandrikes. The quantity so brought in is about 1000 MT p.a. and is less significant in comparison with production in Karnataka and supply from private lands in Konkan.

Use of Bamboo

Bamboo is used as common household material in agriculture, horticulture, sericulture, housing, packaging, transport, storage etc. Industrial uses of bamboo consist of pulping, matboards and as core-sticks in Agarbathis (scented sticks). Generally, green and straight bamboos are used in the domestic sector and in weaving. Slightly defective, bent and dry bamboos are used for pulping. Weaving and core sticks of agarbathis are possible from green bamboos or after soaking dry bamboos.

Dry bamboos are used for strengthening the back of mats in Chandrikes and splits are used for similar purpose in baskets, mats, trays and Chandrikes. Dry bamboos are used to construct frames to support a series of trays in silkworm rearing. The average estimated uses of bamboos p.a. from recorded sources in the domestic and industrial sectors in South India are given below:

	Pulping MT	Other Uses MT
Karnataka	40,000	30,000
Andhra Pradesh	2,10,000	60,000
Tamilnadu	10,000	NA
Kerala	60,000	25,000
Goa	–	10,000
Pondicherry	–	–

Bamboo weaving is a traditional centuries old cottage industry. A large section of weavers belong to a defined community in South India and are termed as Buruds'. Karnataka alone has a population of 50,300 Burud families and AP has 2,900 families. In addition to Buruds, weaving of bamboos is a domestic profession among several families in the low income group of rural communities. Since villagers situated in the vicinity of natural bamboo forests enjoy the privilege of bringing headloads of bamboos for their weaving profession, there is no exact record of quantities removed on headloads.

Woven bamboo is used in several sectors of which sericulture is quite important and compares with use in agricultural sector in South India. Silk industry in South India is a large user of bamboo next only to the pulping sector. Since weaving is an art learnt by rural poor communities through generations, the employment generation in this type of use is much higher than bamboo use as raw material for pulp and paper industry.

The society constituted a core team for initiating the study. Apart from the office bearers of the society consisting of the Chairman, Executive Director and the

Treasurer, all of whom are experienced foresters, an economist who had previous exposure to practical sericulture was taken into the team for advising on statistical and economic approach to the field study for gathering data through a set of designed questionnaires. These data forms were then handed over to a number of field officers to visit selected samples of mulberry farms, silkworm rearing farmers, bamboo equipment manufacturers, intensive worm rearing villages, cocoon markets, silk filatures and reeling centres etc.

Baseline data for sampling was gathered through the Department of Sericulture in the States of Karnataka, Andhra Pradesh and Tamilnadu. However, the sericulture data in Kerala state had to be obtained from the Office of Sericulture Federation (SERIFED) based in Trivandrum. Statewise silk production data during the last five years was readily available from the Office of Central Silk Board in Bangalore. Data on extent of mulberry cultivation, egg producing centers, quantity of disease free layings (DFL) used raw silk, produced etc. was also available from the board.

Districtwise data on extent of silk production and mulberry cultivation was obtained from Directorate of Sericulture in the three states and from SERIFED in Kerala. Data on bamboo resources, harvest and sale were obtained both from the forest department statistics, from bamboo cooperatives, traders etc. List of manufacturers of bulk equipment using bamboos so procured was obtained from the district officers in the department of sericulture, district offices of the forest department and from traders.

Using the baseline data so obtained, it was possible to draw stratified samples of mulberry farmers, bamboo harvesters, purchasers and the groups of artisans engaged in the manufacture of bamboo articles for sale to users.

The mulberry farmers, officials of the sericulture department, manufacturers, owners of silk filatures etc. were then visited for interaction to gather information on the whole chain of production and supply of bamboo articles, costs, extent of use of each product, its average life, cost of replacements, sources from which the materials are either hired or purchased etc. The data so gathered was collated to arrive at averages for requirements of products, their life, costs, socio-economics of production of bamboo articles and cocoons etc.

Production Area

It was found that the few districts of Karnataka, Andhra Pradesh and Tamilnadu situated in the South Central part of the deccan plateau form the nucleus of silk production. These districts are in the tropical zone with advantage of annual rainfall of about 750 mm spreadover 7-8 months in the year and have tropical red loam as the main soil type. This zone has the advantage of both South West and North East monsoons bringing pre-monsoon showers in April-May and precipitation between June and November.

The area enjoys light rains, long sunshine and mild temperature ranging between 15°C in Dec-Jan and touching 34°C in midsummer of May-June. Mulberry farming is sustained by light irrigation to supplement the number of rainy days ranging between 100 to 125 between March and November of each year. Ground water gets recharged repeatedly due to prolonged rainy season, though the intensity of rain is low. The red loam assists in percolation and moisture retention to sustain good mulberry crop for 8 to 10 months in the year, which permits leaf harvests of atleast six cycles at average interval of six weeks.

Districts of Kolar, Tumkur, Bangalore, Mysore and Mandya in Karnataka form the core of mulberry farmers with all the neighbouring districts in Karnataka, Andhra Pradesh and Tamilnadu supplementing mulberry and cocoon production. Naturally, filatures, reeling and weaving are situated around the centers of large scale cocoon markets.

The field officers engaged for data collection in the forms prepared by the core team gathered statewide data. The heads of sericulture deptts. in Karnataka and Andhra Pradesh are based in Bangalore and Hyderabad respectively. However, the Directorate of Sericulture in Tamilnadu is based in Salem, which is hardly 200 kms away from Bangalore and it is not based in the State Capital at Chennai (Madras). This in itself is a proof of the zone touching Karnataka being intensive producer of cocoons. Similarly districts of Andhra Pradesh bordering Kolar district in Karnataka are intensive mulberry farming centres.

Silk production in Kerala is too little and hence it is administered by the State Sericulture Federation. Since major part of Kerala state is under the strong influence of South West Monsoon which brings heavy precipitation between June and September, mulberry farming is too little in the state except Palakkad district, which borders Coimbatore district of Tamilnadu. The field officers were then sent to silk farmers, bamboo cooperative societies, district service centres meant to service small scale industries, bamboo market centres, equipment manufacturers, cocoon markets etc. from where data was gathered in the structured proformae.

Exhaustive forms for gathering data through the field officers were prepared so as to gather information to cover on

i) Bamboo sale centres or outlets both in Government and outside government

ii) manufacturers of bamboo equipment for sericulture either in societies or individuals

iii) units providing equipment on hire basis

iv) cocoon rearing profession etc.

The field officers were given copies of the forms to use in field survey so as to gather all physical and socioeconomic data on use of bamboos in sericulture.

Experience has shown that

(a) the production and control of bamboo is with the state forest departments,

(b) bulk of the trade is with erstwhile bamboo extraction contractors and traders who have now joined hands with the buruds, S.C and S.T groups to participate in one or more cooperative venture

(c) these contractors or traders control the production and sale of equipment meant for sericulture, including enlisting themselves with the concerned government departments

(d) these traders have established production centres where they engage skilled and unskilled workers for production of equipment

(e) they form bulk suppliers both to

i) service centres created by government

ii) farmers, cocoon producers, reelers etc.

f) they have experience and information on the bamboo extraction policies and programmes of the departments, artisans and groups who can supply components of bamboo to fabricate the equipment like trays, Chandrikes

g) they have knowledge and access to bamboo market, mat suppliers etc. from all over India, which they use to keep up their trade and thus, they form the core group controlling production, supply, hire and prices of bamboo equipment.

Since there are many such traders, it is safe to assume that

a) the prices of equipment are competitive enough and vary according to market forces of demand and supply

b) minor variations in the unit cost are controlled by the seasonal variations in the cost of labour

c) quality of the product also reflects in the sale price and

d) the policy and efficiency of the forest department is a major factor governing supplies of raw material and prices.

It was found out that the intervention through the orders of the Hon'ble Supreme Court in 1997-98 and the constraints of forest management due to restraints by the provisions of the Forest Conservation Act, did upset smooth extraction and supply of bamboos mainly in Karnataka. Yet the traders managed to establish contacts with private suppliers in Maharashtra and Goa, mat manufacturers in N.E. India and thus were able to keep up production and supply to the sericulture farmers. There are some schemes of the Government in the forest department, sericulture department, tribal welfare etc. in which there is provision to provide equipment to the rural poor free of cost.

The production of cocoons gets subsidies and governmental assistance at different stages from supply of disease free layings, mulberry farming, supply of equipment to fillip for marketing of cocoons. The establishment of service centres under the Ministry of Small Scale Industries has not changed the mode of production and supply of bamboo equipment to sericulture industry. These district service centres have again listed the bamboo traders/societies to whom they pass on orders of supply of equipment at negotiated prices and keep 2% of the cost of supply as their service charges.

Mulberry Farming

Mulberry farming is an essential part of sericulture. There are not many large farmers cultivating over 4 ha of mulberry. Over 96% of mulberry farming is with small farmers. Most farmers in the traditional silk production zone like to set aside 0.25 to 0.5 ha out of their average holding of 1 to 4 ha for mulberry cultivation. In over 90% of such farmers, the work of harvesting leaves and rearing of silkworms is done through the family members, because it engages most of their idle hours. Since the work of harvesting leaves, chopping them, feeding to larvae, transfer of larvae from trays to trays or finally to Chandrike, is done by family members, without distinction between age, sex, physical or mental skills etc.

The cocoons gathered from Chandrikes are carried on headload, cartload, buses or stage carriers for sale in the government supported cocoon market or even private traders. The business of handling cocoons and separating silk thread and waste after steaming the cocoons is a sticky and smelly one. This is done in filatures which are a domestic or tiny scale industry, traditionally run mostly by muslims. Here again the whole family and some casual local labour is involved.

The department of sericulture has maintained annual statistics of area under mulberry cultivation, number of families and villages involved in mulberry cultivation, the quantities of disease free layings (DFL) supplied for rearing of worms, quality and quantity of cocoon production and production of silk and silk waste in the states under study. Extract of such data obtained from Central Silk Board is summarised in the following table as averages for five years from 1995-96 to 1999-2000 for the four states studied. Pondicherry and Goa in South India donot have records of sericulture.

Sl. No	State	Area under Mulberry - ha	DFL Nos 00,000	Reeling Cocoons tonne	Raw silk tonne	Silk waste
1	Karnataka	1,57,108	2017.11	73,572	8578.4	2586.8
2	Andhra Pradesh	38,744	576.58	26,686	3018.6	782.02
3	Tamilnadu	11,578	184.03	6,805.4	777.6	215.4
4	Kerala	1,359	1.12	33.65	3.09	0.68
	Total	2,08,789	2778.84	1,07,097	12377.69	3584.88

It is found out that the districts having large mulberry cultivation are:

— Bangalore, Kolar, Mysore & Mandya districts which account for 77.3% in Karnataka

— Anantpur, Chittoor, Kurnool and Cuddapah which account for 85% in A.P.

— Dharmapuri, Salem, Erode and Vaniyambadi districts in T.N. - 70% in Tamilnadu

— Cannonore, Idukki and Palakkad districts in Kerala - 70% in Kerala.

The average number of families holding sericulture farms in these states is summarised below:

	Total Families	Av. Holding Ha.
Karnataka	3,33,823	0.47
Andhra Pradesh	51,724	0.75
Tamilnadu	37,450	0.31
Kerala	19,651 (0.69)	0.07
Total	4,42,647	0.471 ha

Usages in Sericulture

Bamboo thorns and splits are used for fencing in farms. But this practice is limited to natural bamboo zones, which are away from sericulture farms. No assessment is made for bamboo thorns use as it is common practice only in high rainfall where paddy & sugarcane are cultivated.

Baskets

Two kinds of baskets are used in mulberry farming. One basket is for manuring the crop. The second basket is for gathering leaves. The manuring basket is about 0.6 mt diameter and about 25 cms deep, circular in shape. The leaf collection basket has 55 cms middle diameter tapering to about 20 cms at the bottom and 20 cms mouth at the top. Average weight of both the kinds of baskets is 1.5 kg each and have average life of three years. Average market price is Rs.30/- each.

The number of baskets needed is worked out by field studies at 10 per ha for application of manure and six per ha for leaf collection. About 80% of farms use bamboo baskets. The manure basket can hold 8 to 10 kgs of manure and leaf basket can hold 12 kg of fresh mulberry leaves. The baskets are woven with 2 to 3 cms wide bamboo slivers strengthened by bamboo strips of 2 to 3 cms thickness. About 80% of farms use bamboo baskets. Leaf harvest is done upto nine times each year. Quantity of leaves harvested is nearly 35 tonne p.ha p.a. for irrigated mulberry and half this quantity for rainfed crop. Based on the above averages, the requirement of bamboo of these baskets is worked out for the 4 states and given in the table below:

1	Area under mulberry in the southern states	2,08,789 ha
2	At 80% of farmers using baskets	1,67,030 ha
3	Baskets at 10 per ha for manuring	1.67 million
4	Number of baskets divided by average life of 3 years	0.55676"
5	Weight of 1.5 kg each - tonne	8.35
6	No. of leaf collection baskets at 6/ha - million	1.002
7	No. divided by 3 years life - million	0.334060
8	Weight of baskets at 1.5 kg each	501 tonne
9	Weight of bamboo required at 10% wastage 835+501	1469 tonne
10	Manpower generation at 5 baskets per day	0.385
11	Wage earning at Rs.50/- per day	Rs.19.25 million
12	Market value of the baskets at Rs.30/- each	Rs.26.72 million

Mats

Bamboo mats are regularly woven by the traditional weavers for various end uses. The most common uses are for partitions, fencing for cash crops like betel-leaves, dunnage in stores, for spreading in sun drying of sugar, grains etc. In addition, the mats are used for chopping mulberry leaves, storing steamed cocoons before drying them in cocoon storage trays kept on stands, one above the other. Normal size of woven mat is 1.8 mt x 1.2 mtrs. Average airdry weight of a mat is 1.5 kg. The weaver sells such mats at Rs.22/- to Rs.25/- each. The market price is Rs.35/- each. Mats are rolled into bundles of ten and sold in the market in South India.

Such mats from Assam come in bundles of 35 in each roll. About 3500 mats are transported over the long distance of over 2000 kms to consumption centers in South India. Assam mats are superior. They are sold in retail market at Rs.45/- each and are mainly used for fabricating mountages. Average life of a mat is two years. Farmers' family uses 5 mats per annum.

Mats in Filature

After steaming the cocoons in the stifling basket, the hot soft cocoons are unloaded on a bamboo mat. Average number of mats used in equal to the number of stifling baskets in use. The mat has average life of two years due to contact with heat and rough usage. At 10 kgs of cocoons per basket as capacity of holding and at 80% filatures using baskets, the requirement of mats and baskets is calculated. Production of reeling cocoons in the four states under study is 108,380 tonne per annum.

At one mat for 10 kg of cocoons, the requirement of mats is 0.86704 million at 80% of farmers using such mats. Mats in filature are used for average life of 200 times. Each mat is 1.5 kg in weight and market value is Rs.25/- each.

Rearing Trays

Woven bamboo trays, generally in round form are used for feeding of worms. The trays are generally in size of 1 mt and 1.2 mt diameter with a rim of 6-7 cms height. The requirement of such trays increases as the worms grow in size after every moult. The bamboo trays called "Tatte" are vertically placed one above the other on frames which are made of wood or bamboos. Generally, the outer frame is that of wood and the partitions are either wooden reepers or round bamboos.

Each frame is generally 2 mtrs in height and can hold 20 trays in two rows. This is meant to save rearing space in the village homes. Either a part of dwelling house is used for accomodating such frames or rearing houses protected from ants are specially constructed for such use in zones where cocoon production is a repeating annual feature. Each tray has weight from 2.0 to 3.5 kg, with an average dry weight of 3.0 kg. Trays are woven from 1.5 to 2 cm wide slivers of both Medri and Dowga bamboos and strengthened by splits of 1.5 to 3 cms.

The rim of trays require special strengthening by thick splits. Bamboo trays are not used in case of branch feeding which is resorted to by some large farmers to save on cost of labour required in leaf harvest, chopping and feeding. Branch feeding is done on large frames of size 3.5x1.5 mt made of wood or bamboo frames, on which coir ropes or plastic ropes are tied to hold the branches. From field studies, it is estimated that 82% of farmers use bamboo trays. Number of leaf crop of mulberry harvested varies from 5 to 9 p.a. depending on moisture and conditions of growth.

Average number of crops is 8 in the traditional zones under irrigated cultivation and five under rainfed cultivation. The trays are used repeatedly and have an average life of 3 years. The trays are smeared with mud-cowdung mixture and are kept disinfected by spraying insecticides. Requirement of number of trays has been assessed by contacting in any samples of farmers. Number of rearing trays required per annum and related data is given below:

Area under mulberry in the southern states	2,03,789 ha
Requirement of trays at 200 per ha	41.7578 million
Number at average life of 3 years	13.9192 million
No. reduced to 82% usage	11.1354 million

Weight of trays p.a. at 3 kg per tray	33406 tonne
Weight of bamboos required at 10% wastage	36746 tonne
Value of trays at Rs.50/- each	556.77
Employment generation at 4 trays/day/pair	5.567
Wage earnings at Rs.100/- per pair per day	278.38

Sources of Supply

There are some manufacturers and traders in traditional mulberry zone who are major suppliers of trays. The manufacture of trays is done by artisan families for which bamboos are procured mainly from government sources and supplemented by private purchases. One big bamboo (Dowga) can give 2 trays. Conversely, two thin medar bamboos may be needed per tray. The circular portion in the form of mats is sometimes made in one place and the trays are fabricated near the farm centres like Chintamani in Kolar dist., Vijayapura and Kanakapura in Bangalore rural district, Maddur and Malavalli in Mandya district, Sira in Tumkur district etc.

Average of four trays are made per day for which the total value is Rs.200/- and the wages earned by a pair of workers is Rs.100/- to Rs.120/- per day. The market price varies from Rs.40/- to Rs.60/- each depending on size, quality and season.

Mountage (Chandrike)

This equipment is of great importance in cocoon production. Mature larvae which are about 23 days old are transferred to the mountages in the circular thin frames between which cocoons are spun. The spinning process takes 48 to 60 hours after which the cocoons are removed for sale. The remnant waste silk is removed, the mountages cleaned and disinfected before reuse. A mountage is made of 1.8 mt x 1.2 mt mat strengthened with solid bamboo at the back by tying with coir rope.

On the front side of the mat, 45 to 50 mtr long woven bamboo ribs of 5-6 cms high are fixed concentrically by coir rope. These are called "flowers" (Hoovu in Kannada) and are either locally prepared or brought from bamboo production centres like Shimoga and Nandyal. The mats are also locally woven or bought in bundles from Shimoga and Gauhati in Karnataka and Assam respectively. Though Thrissur and Trivandrum in Kerala manufacture large quantities of mats of reed bamboo, these are not used in sericulture, though the prices are competitive.

Kerala State Bamboo Corporation is the main manufacturer of mats through many cooperative societies in Kerala by supply of reed bamboo given free of royalty to the corporation by Kerala Govt. A mountage weighs from 10 to 20 kg depending on the weight of bambos used. Average weight of mountage is 13 kg of which 12.5 kg is the weight of bamboo and the balance is that of coir rope.

Field studies have assessed requirement of mountages per ha, per crop and for every 100 dfl. 80% of farmers use bamboo mountages, which are priced at Rs.160 to Rs.250/- each depending on quality and season. They are also kept repaired if slightly damaged. Mountages are bought by large farmers and are often hired during peak production for 3 days of pupation. Hire charges are Rs.5/- to Rs.6/- each per use. It is revealed that 40% farmers in Andhra Pradesh own mountages and 60% hire them. Average life of a mountage is three years even after some repairs. Research studies have also been conducted on requirement of mountages per 100 dfl or per ha of mulberry cultivation which is Rs.250 per ha.

Labour and value: Two mountages are made per day if the whole manufacture inlcuding splitting, slivering, weaving and fabrication are done at the place of manufacture. But mats and flowers are bought in bulk and fabricated at the manufacturing place in which case 5-6 mountages are got ready per day by one pair of workers. Average wage earning comes to Rs.40/- per day for ladies and Rs.60/- per day for gents respectively. Cost of bamboos comes to an average of Rs.6/- per mtr length. Mountages used:

Area under mulberry in the 4 states	2,08,789 ha
At 80% farms using mountages	1,67,031 ha
Mountages required at 250 per ha	41.7578 million
Divided by average life of 4 years - p.a.	10.439 million
At average wt. of 11 kg per mountage	1,14,830 tonne
At average price of Rs.200/- each market value	2087.8 million
Bamboo used at wastage of 5%	1,20,571 tonne
Employment at 3 mountages per day per pair of weavers	6.9593
wages earned at Rs.50/- per weaver per day	347.965

Cocoon Storage Trays

These are like rearing trays woven from slivers and splits of bamboo. The trays are one meter in diamater and 5 cms in depth. Average number of trays used for storing cocoons produced from 100 dfl is 15. The trays are priced in the market at Rs.40/- each. Average weight of the tray is 2.5 kg and it can hold 12.5 kgs of cocoons. The average life of these trays is five years.

The requirement of cocoon storage trays is worked out on average Dfl. Used each year in the states under study and calculated for only 80% of cocoon production, as wooden trays are also used. One pair of weavers can prepare five trays per day. Two medri bamboos are needed per tray.

Steam Stifling Baskets

Most of conversion of cocoons to silk is carried out in the filatures on a cottage industry scale. The filatures are around the cocoon markets. A filature processes about 300 kg cocoons per day and works for about 250 days in a year. Average production of cocoons in the four states is 108,380 tonnes p.a. 10 kgs of cocoon are stifled in steam per batch in a basket.

The requirement of baskets p.a. comes to 43,352 p.a. on the basis of 200 times usage for each basket. Since the average weight per basket is one kg, the total weight of baskets is 43.352 tonne or roughly 45 tonne of airdry bamboo. At six baskets woven per day per weaver, the man power used is 7,225 mandays per annum. The value of baskets at Rs.20/- each comes to Rs.0.867 million p.a. The wages earned at Rs.50/- per day comes to Rs.0.722 million p.a. Requirement of stifling baskets:

Quantity of cocoon production p.a. in southern states	1,08,380 tonne
No. of baskets at 10 kg capacity at 80% usage	0.86704 million
No. of baskets p.a. at average of 200 uses	0.0043352 million
Wt. of baskets at 1 kg each	43.35 tonne
With allowance for wastage bamboo need	45.00 tonne
Market value of baskets at Rs.15/- each	Rs.0.650 million
Manpower requirement at 8 baskets per day	5420 mandays
Wages earned at Rs.50/- per day	Rs.0.361 million

Bamboo Sticks in Rearing Stand

The trays are placed one above the other in rearing stands. Ten trays in two rows are placed on such stands in the village houses to save space. It is seen that the frames are made of jungle-wood, but bamboo sticks are used as partition in 80% of such stands. The sticks are of round bamboo of length of 1.5 mts each. Each stand needs 20 sticks, which can be obtained from 5 bamboos. The average life of sticks is three years. As 200 rearing trays are used per ha p.a., the number of stands per ha is 10. The requirement of bamboo for construction of stands is worked out and given below:

Bamboo sticks in rearing stands

Area under mulberry in 4 states	2,08,789 ha
At 10 stands per ha No. of stands	2.879 million
Considering 80% usage, No. of stands	2.3032 million
No. of sticks at 20 per stand	46.064 million
At four years average life	11.1353 million
No. of bamboos at 5 sticks per bamboo	2.227 million
Weight of 200 per tonne bamboos	11,135 tonne
Value of bamboos at Rs.20/- each	Rs.44.54 million
Manpower at 200 sticks per day	0.0557 million
Wages at Rs.40/- per day	Rs.222.7 million

Chawki Sticks

The requirement of chawki sticks is small. Only 5 sticks are used per ha. The sticks have average life of 3 years. Since each stick is about 1 mt long and 3 to 4 cms in diameter, the number of bamboos at 5 sticks per bamboo works out to bamboo numbers equal to the area of mulberry per annum divided by 3 years of average life. The bamboo requirement is given below:

Area ha	No. of bamboo needed	Wt. In tonne at 200bamboos/tonne	Rs. Million
2,08,789	69,600	348	1.392

Other Usages

Besides the direct uses in sericulture, there are many

other uses connected with harvesters and weavers. The bamboo felling labour has to stay in temporary sheds constructed from use of bamboos, mats and leaves. The bamboo felling operation in South India takes place between December of each year upto end of May of the next year. Fellings are suspended during the rainy season between June and November. Since the bamboos are extracted for various end-uses, including for pulp mills, the use of bamboos for temporary dwellings cannot be accurately assessed for sericulture end use. Residues of fellings are available as fuelwood for such labour in the forest.

Sheds for Weavers

Most large-scale manufacturers create bamboo-made workshops for storage of bamboos, shelter to workmen, storage of articles manufactured etc. Such sheds are used for weaving of other bamboo products like vegetable baskets, agricultural needs like grain storage bins, winnows etc. Hence it is not possible to assess bamboo requirement exclusively connected to silk industry.

Employment

Extraction of bamboos in the forest from thorny intertwined branches in the clumps is both difficult and time consuming. Even the approach to the base of clumps requires clearing of basal thorny branches. Evenafter cutting the base with a long handled axe or bill-hook, the culms have their branches intertwined with others at the middle and top of clumps. The current requirement of labour for bamboo extraction is worked out on averages now used. The felling contract is given on the basis of 100 bamboos in case of better culms used in weaving and tonnage or volume in case of industrial uses.

In addition, labour is used for loading, unloading, storage etc. The present average requirement for felling and extraction of bamboos from forests consists of six mandays per tonne Loading, unloading and stacking require one manday per tonne. Average wage rate now is Rs.75/- day.

Economic Importance

After the silk threads or silk waste is produced, there are several stages upto the use of silk which has great significance in the gross national product and employment generation. Spinning, colouring, blending, weaving of silk into fabric or sarees and manufacture of apparel are all the downstream activities. Bulk of weaving is in the rural sector through handlooms or powerlooms. Though some components in such weaving can be of bamboos, no further study for such uses is deemed necessary since wood, plastics and metals have been progressively used in place of bamboo.

Cardboard trays, plastic trays and mountages have been devised and introduced in silkworm rearing. But the traditional cocoon rearing farmers in South India prefer bamboo trays and mountages in preference to other equipment. During the field sampling, almost all farmers have been unanimous in their opinion that bamboo equipment is still the best and have suggested removal of bottlenecks in the supply of bamboos, easy movement, standardisation of products and costs, rather than find substitutes.

Problems in Supply

Almost the entire supply of bamboos as raw-material for manufacture of equipment comes from Govt. forests. Bamboos are defined as "forest products", "whether brought in from the forests or not" under the forest act. Therefore, transportation of bamboos has to be covered by a permit duly issued by an authorised forest officer. Procurement of such transit permit involves delay in movement of harvested bamboo, resulting in losses due to driage, pilferage and decay due to attack by fungi, insects etc.

During the last 10 years, harvesting of bamboo has been taken over by the forest department. This has introduced more red-tape, delays and some corruption too, in addition to higher costs of extraction, which has to be borne by users. Weavers of mats, trays and manufacturers of mountages have to put up with these problems. Billhooks tied to long bamboo poles used to cut the obstructive branches higher up. The culm is then dragged by holding the buttend either by a set of labourers, by hand or by tying a rope around the buttend.

The felling operation has become cumbersome because of unscientific harvests, resulting in congestion of clumps and intertwining of culms through the clump while attaining heights. Heavy grazing compacts the forest soil, which is also responsible for congestion of clumps. These factors can be remedied, if the bamboo forests are scientifically managed. The weavers, traders and other artisans have to accept the quality, quantity, age of culms etc. as handed over to them, without selection. Certain percentage of material sold to them cannot be used by them.

Weavers would like green and soft bamboos for ease of weaving, quality of product and least wastage through residues. Since users have no control on selection, harvest and transportation of bamboos, they suffer both in quality and cost of the products. There are instances when the supply of bamboo is made long after their felling, resulting in driage, decay and total loss of the produce for use in sericulture. The situation is no better for extraction by the forest development corporations.

It is an irony that there is neither improvement in quality or quantity of bamboo production in all the states, even after introduction of departmental harvest. There has been no attempt at introduction of silvicultural management of clumps or scientific harvest. No modernisation in tools and methods of harvest, transport, storage, grading and no preservative treatment is resorted to. The policy of governments at concessional supplies to artisans and their societies has been misused for an unhealthy nexus between politicians, officials and traders. Traders backed by the politicians control the societies.

Even the ministry of textiles, sericulture and the departments of sericulture aswelas Central Silk Board, have been trying to find substitutes to traditional bamboo equipment and not given adequate thought to improving quality of bamboos or smooth and adequate supply of bamboos, which traditional farmers would like to use in preference to substitutes. Only in cases of non-availability and unreasonable high prices, the farmers use the substitutes like cardboard and plastics.

Branch feeding is resorted to by large farmers to save on cost of labour and not for better results. The agencies promoting mulberry cultivation and sericulture have not looked into improvement in supply of much needed bamboo equipment. A suggestion made to introduce useful bamboo species as border-crop around mulberry farms, has also no takers.

Bamboo Management

Extraction of bamboos from forests was prescribed as an overlapping working circle on a 3 to 4 years cycle in the working plans drawn for forests. Rigid conservation policies, reservation of bamboo areas as wildlife preserves, stoppage of green fellings due to lobbying by ill-informed environmental activists, diversion of attention of forest staff to social forestry resulting in neglect of normal forestry operations, interventions of the Hon'ble Courts resulting in lengthening of the red-tape in approval of management schemes etc. have all resulted in irregularity of extraction and supplies.

Periodic flowering and death of clumps worsens the situation. It can be said that the famine of bamboos in the market and related miseries to artisans and users, can be attributed to lack or neglect of forest management and more so in the last five years. Even the bamboo plantations raised under different plan schemes have remained neglected, resulting in very poor yields and no improvement in yields. The table below shows data of extractions of bamboo in Karnataka to illustrate fall in yields. The same trend is seen in Andhra Pradesh and Tamilnadu.

Years State	Period 1976-80	1980-1985	1985-1990	1990-1995	1995-2000
Av.Karnataka Qty.	1,01,507	1,17,192	48,948	25,169	21,882
TonneA.P.	NA	13,29,255	2,36,090	NA	1,70,000
p.a. T.N.	—	17,461	6,635	4,010	1,500

Due to neglect of management, apathy of Govt. departments to the needs of the artisans, legal and administrative hurdles in extraction and supply, the artisans and manufacturers of equipment suffer from uncertainty, irregularity of supplies, fluctuation in prices etc., which in turn cause problems to workers, sellers and users.

The fall in yields from forest extraction and obstacles in easy availability have resulted in homesteads in Kerala and Goa grow bamboos due to ease of cultivation and ready market. Konkan districts of Maharashtra along the border of Karnataka have also taken up cultivation of bamboos. Species grown in Kerala is Bambusa bambos. In Goa and Maharashtra, it is Pseudoxytenanthera stocksii in bulk and medri in smaller numbers.

An estimated annual production of 10,000 tonne is recorded from private sources in Kerala and about 30,000 tonne between Goa and Maharashtra. Tanjore district in Tamilnadu has taken up bamboo cultivation in private lands, but the produce is locally used as props for banana cultivation. The artisan groups dealing in equipment for sericulture buy private bamboos in bulk at markets in Sankeshwar and Belgaum and transport them to destinations of drier districts in Southern Karnataka, neighbouring Andhra Pradesh and Tamilnadu for their use. They have managed to solve problems of permits, checkposts and transport obstacles on their own at some additional cost, which is a fraction of the cost of purchase and transport.

Manufacturing at Site

Except for the wholesale contractors who number 3 to 15 in different towns in Karnataka and bordering A.P. and Tamilnadu, hundreds of artisans live in hutments. There are lanes in the towns where the buruds are concentrated. Each household either buys a headload of bamboos numbering 20 to 30 depending on the size at Rs.30/- and the family members do the splitting and weaving. There are many instances, where the wholesaler supplies the raw material and pays wages fixed for the products made to order.

Some villagers situated near bamboo forests collect bamboo on headloads with or without permit and sell products to wholesalers. Individual families are always needy due to irregular availability of raw material because of which, they have idle hours and days or due to time required to get back the value of bamboos and their maintenance costs, which comes after sale of the produce. This is one reason for which they either depend on advance amount obtained from traders for which they commit quantity and rate for supply or depend on the traders for supply of raw material.

The larger manufacturers or traders have been in this business for 15 to 30 years and know the tricks of the trade. They also have knowledge on alternative sources of supply of raw material or semi finished products used in fabrication of trays and mountages at different centres in Karnataka, Andhra Pradesh and even Assam for supply of mats. They purchase bamboos, mats, flowers (used inside Chandrike), coir-ropes etc. in bulk and supply the same to individual families for manufacture or hire artisans on wage per unit of manufacture.

In the latter case, they construct large shed or pandal mostly out of bamboos, mats and plastic sheets. These sheds and pandals are used for storage of raw-material upto one or two lorry loads and storage of finished products till they are sold. Weaving and fabrication is done either outside or inside these pandals.

The cost of purchase of mats of size 1.8 x 1.2 mts varies from Rs.25/- to Rs. 35/- in wholesale and Rs.40-45 in retail. Bamboos are bought from forest department sources, bamboo cooperative societies, forest development corporations or bamboo merchants in Belgaum district, Shimoga, Rajamundry, Kurnool, etc. Flowers of about 45 m length are bought at Rs.65 to Rs.80/- each and its length is tailored to fabrication of one mountage.

Coir rope bundles come from Kerala or Coastal Karnataka. It is seen from field study that the experienced traders succeed because of their skills to overcome problems in procurement of raw material and their ability to make investments upto about Rs.2,00,000 during peak season to sustain their trade. In cases where there is lack of ready demand for mountages for sale, they hire them to small farmers at Rs.5/- to Rs.6/- for a 3 days requirement and keep recycling them.

In Karnataka about 40% of rearing farmers own the mountages and 60% hire them. Those depending on hired mountages number 75% in Andhra Pradesh and Tamilnadu. The annual production of larger bamboo traders varies from 3000 to 10,000 trays and 2000 to 6000 mountages per annum. Trays are generally bought and not hired. But many small farmers take both mountages and trays on loan from neighbours or on barter to avoid purchase.

The department of sericulture should promote bamboo plantations along with mulberry in the zones where artisans need supply of fresh bamboos. Such bamboo planting can be in homesteads, boundaries of mulberry fields or even block planting in common property resources. Service societies should be encouraged to motivate farmers to grow bamboos on assurance of purchase at support price, easy credit facilities, supply of quality planting stock, technical and extension services. Such societies should help in bamboo planting, harvest, contact with agencies having technical competence in bamboo planting, harvest, preservative treatment and market information.

User groups should be attracted by the forest department and the forest development corporations in joint forest management of extended bamboo plantations for mutual benefit. Information technology should be harnessed for dissemination of market information between producers, traders, processors and users for proper linkages. Forest and sericulture research institutes should undertake applied research in continuous improvement of equipment in usage to reduce on cost of purchase, maintenance and replacement. Manufacture and supply of improved tools for splitting, slivering, thicknessing and joinery by service societies can help in reducing labour costs, wastages and improve quality of the product.

Storage of bamboo equipment during idle period has not been given attention and deserves educating the users on the same. Promotion of a multiplicity of products and their ready sales for reasonable returns to the artisans, will help in reduction of wastages of bamboo, avoidance of idle time and improved wages to the poor families of artisans. The department of small scale industries and the handicrafts development board should have special cells at the district level to study, identify and remove bottlenecks in the chain of activity from resource to the marketing of the products.

Help and advise from voluntary agencies engaged in bamboo development should be continuously taken by the producers of bamboo like forest department, forest corporations, joint forest management committees etc. on the one hand and the service centres and wings of development departments on the other, for improving coordination.

5

Biology of Silk Production

Spiders invented silk circa 400 million years ago. During the last 180 million years, increasingly complex silks have been employed in the advanced web engineering necessary to cope with the ever more demanding specifications required in the aerial arms race with jumping and flying prey. In addition to being major construction materials in webs, silk fibers are used by spiders for retreats, trip lines, traps, brood chambers, and safety draglines. Thus, spiders, especially the advanced web spiders, have evolved veritable armories of highly specialised silks (Figure 1). Yet man has never successfully exploited spider silk, mainly because the cannibalistic nature of spiders does not lend itself to high-density husbandry.

Although insects also have found many ways of making silk, web spiders are unique in maintaining—in the same individual—a battery of highly specialised glands producing very different silks, often simultaneously. This diversity is driven by the central role played by silk in all aspects of a spider's life and by the close co-evolution of the material, of the spider's behaviour in using it, and of the prey's behaviour (in trap silk) in avoiding or breaking it.

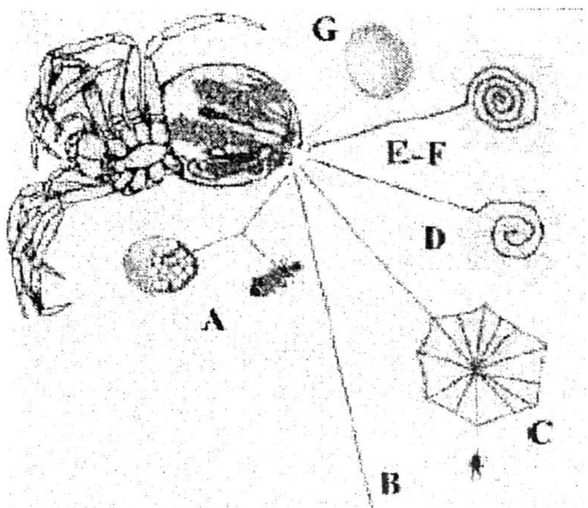

Figure 1. The female cross spider Araneus (like most other advanced orb-weaving spiders) spins 7 different silks from as many different glands: (A) Soft silk, to enshroud prey, cradle the eggs in the cocoon or to balloon. (B) Cement silk to affix the web to support. (C) Strong but relatively stiff silk for the scaffolding of the web. (D) Additional scaffolding silk also used for the temporary support spiral. Very elastic sticky capture silk consisting of a core thread (E) and a watery coat (F) that forms the web's droplets and glue. (G) Tough silk for the egg sac "shell".

There is another reason for focusing on spiders—their natural ability to produce silk at all times as well as copiously. Most insects with commercially interesting silks, such as the silkworm Bombyx mori, produce silk only once in their lives when building a cocoon and resist being manipulated during that time. Spiders, on the other hand, when immobilised, find it impossible to cut a thread that it is being pulled from the spinneret. And thus the spider can be silked day after day for weeks on end.

Coupled with the outstanding properties of specific draglines, this experimental access, probably more than any other feature, has led in the last few years to many novel insights into silk spinning in general. Admittedly, the rapid rise of novel analysis techniques that can be applied on single fibers as they are spun from the live animal under controlled conditions, such as noninvasive micro X-ray diffraction or Raman spectroscopy, was crucial for many of the studies.

Also important were advances in sample handling and the design of dedicated stress-strain gauges. After all, a typical dragline thread is rather thin, at around 2 μm in diameter, while a typical silkworm thread is much thicker, consisting of two brins of 10 μm in a bave of about 30 μm.

Historical Outlook

For nearly 4000 years, silkworm silk was a mainstay of commerce as one of the most expensive of luxury goods, bringing wealth to traders and their communities as well as employment to the producers. The decline of traditional dress in the early years of the 20th century, coupled with the demands as well as the constraints of the Second World War, helped petrochemical fibers to gain their present strong position at the cost of natural silks. It appears that silks, albeit now artificially made, are on the verge of a comeback. But to successfully copy natural silks, we must mimic the salient design features of both the feedstock dope proteins and the spinning process.

Molecular biology has led to the extraction, synthesis, and assembly of gene constructs as well as their expression in animal and plant production systems, which in turn should soon be able to supply feedstocks for spinning. Classical morphology provides first construction details of the silkworm's and the spider's extrusion systems.Together, these two biological disciplines have provided the platform that will lead to the designs of prototypes for biomimetic extrusion technologies. Such biomimetic spinnerets should have ecological appeal, as they, like the spider, will be using aqueous protein solutions for low-cost extrusion of exceptionally tough or extensible fibers.

Spider Silks

The different silks in an advanced web spider like Nephila probably evolved from a type of gland that would have been a simple sac with spigot (Figure 2b). Such glands typically are (still) the main glands of ancestral spiders such as the mygalomorph *tarantulas*.

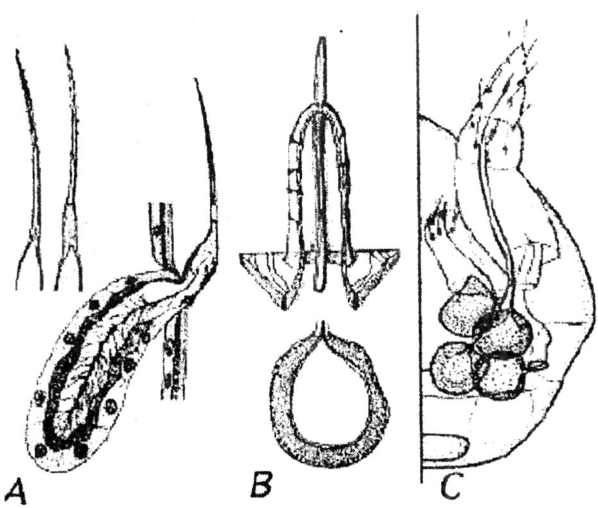

Figure 2. Evolution of silk glands in spiders. (a) Ancestral "spinning" glands were simple sacs with hairs that had to break off to release the spinning solution. (b) More advanced spiders had larger, often more complex glands that exited through a duct opening at the tip of a more or less mobile turret. (c) Specialisation led to multiplication of glands as well as growing size. This required longer ducts to the exit, which in turn led to increased complexity in the spinning process after.

Relatively simple glands (probably little changed over hundreds of millions of years) are also present in large numbers in advanced spiders. Here they produce, for example, the very fine silk that is used in wrapping prey. It can be imagined that the requirements of making thicker silk, or making longer threads, have led to the evolution of bigger glands.

The problem of housing these glands in the long, thin spinnerets, coupled with constraints of making the spinnerets thicker (which in turn would constrain their agility), would then have led to moving the glands up into the abdomen. The gland's moving away from the exit of the body would automatically have led to a redesign of the duct itself to handle the now much longer passageway. In turn, this would lead to a redesign of the feedstock to facilitate this now rather different rheology in the evolved and modified duct.

Evolution is rather good at quick adaptation of specific design parameters, especially if the selection pressure is strong. The need to accumulate nutrients, which directly translates into growth and reproductive fitness, is a strong selective force. Unlike industrial design, which typically progresses sequentially and ideally using logic, biological design progresses at the same time both in parallel and in sequence, using the "blind" trial-and-error approach of random mutation coupled with some type of functional consolidation. This can lead to very rapid evolution indeed, although it typically prevents a major single-step modification of the design. Instead, tinkering is the rule, with add-ons to ensure that things work at all times. After all, the animal (or plant) can never shut down for a total "renovation", and "rethinking" an obsolete design parameter is out of the question.

Design Requirements

Spiders use silk for many purposes, of which prey capture is the most interesting. The mechanical functions of many of the other silks are less easily understood; moreover, their properties are—at present—less in demand for potential human use. On the other hand, fibers with the qualities of web silks seem to be of great interest. Webs, with the orb-web as the most striking example, typically is an aerial structure that has to absorb the kinetic energy of the prey arriving at some speed. Clearly, web design requires rather specific mechanical criteria for the material properties of the silks employed.

Thus, the typical orb web incorporates three principal silks: one silk (of the threads of the stays and radials) is a rather stiff material with extensibility of under 40%; one silk (of the capture threads) is much softer, with extensibilities of over 400%; and one silk (the cement that affixes the capture threads to the radials) is stringy, and its properties are not known. The stiff silks of the struts and runways are the spider's highways and its communication network, while the soft silks of the sticky capture spiral interlink these struts without impeding their function as vibration detectors.

Silk is a protein and its production requires both building blocks (amino acids) and assembly metabolism, both of which are unable to contribute to the animal's growth and reproduction. Thus, silk carries a production cost that is under strong selection to be optimised; after all, growth and reproduction are the final currency for an animal's success over its competitors and the environment. This optimising selection—balancing and trading off constraints, costs, and benefits—led to a number of silk features such as extreme thinness, recycling (where cost-effective), and liquid crystalline spinning.

A fine diameter minimises the use of protein and at the same time makes it more difficult for the insect prey to see the threads. Recycling allows the spider to recuperate amino acids and trace elements. Liquid crystalline spinning minimises the cost of forming the thread in the first place. At present, only a few spider silks have been examined in any detail, with the major ampullate silk of various Nephila golden orb weaver species being the most prominent.

Spinning Micro-and Nanocomposites

The performance of an orb web, which is designed by evolution to take out-of-plane load in maximum deflection, incorporates into one web the mechanical properties of several different types of stiff radials and extensible capture threads. While the radial silk is traditional silk (albeit very tough), the capture silks are surprisingly different.

The garden cross spider Araneus, like the golden orb spider Nephila and all other orb weavers of ecribellate spider families, employs in its capture threads a windlass mechanism that allows supreme extensibility at very weak forces, while absorbing the high kinetic energy of the prey without breaking at large stretch.

This intriguing micromechanism relies on water plasticising the core fibers as well as providing surface tension to power the windlasses that roll up into tight balls the structural core fibers stretched by the impact of an insect. These spiders have evolved an aqueous coating, supplied and maintained by hygroscopic compounds that attract the necessary water from the atmosphere.

The "spinning" of these composite fibers relies on simple physics (the instability of a growing water cylinder) and thus constitutes a prime example of self-assembly of a complex fiber (Figure 3). This is possible because the aggregate glands that provide the coating do not exude typical silk fibrous proteins, but rather a mix of compounds that have been sequestered largely from amino acids used in neurotransmission. This mechanism is surprising cheap and allows for high spinning speeds.

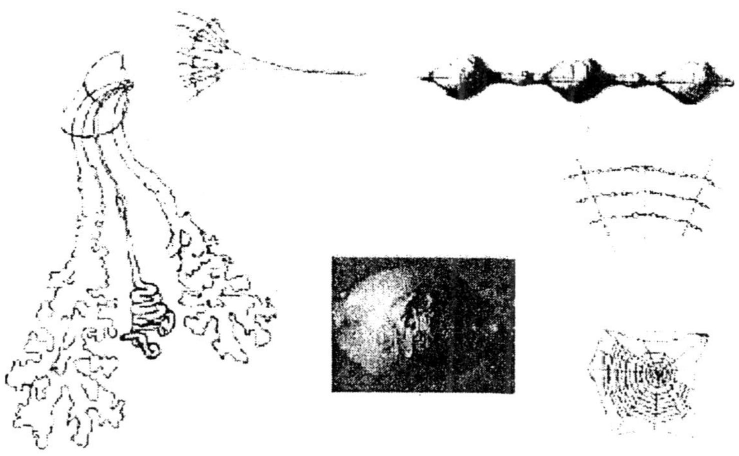

Figure 3. Araneidae orb weavers like the garden cross spider make the capture spiral threads by coating the flagelliform gland core fibers with exudates from two accompanying aggregate glands

Other orb-web spiders, like the hackled band weaver Uloborus, use much more costly ways of making a capture thread with similar (albeit somewhat inferior) mechanical properties. They are much more costly in both silk material and spinning time. Here, the spider combs out its capture silk from a gland providing the strong core fibers, accompanied by a thread from another gland that is crimped to give a coiled spring. Finally, they are covered by a third type of silk, of nanometer diameter, originating in hundreds to thousands of very fine spigots.This latter silk is combed out by the legs to form a loosely puffed up (hackled) fibrillar "wool" covering a twisted rope with a mechanical coil-and-spring that sticks through electrostatic forces.

Spinning Conditions

That Affect Mechanical Properties Spider silks are highly variable not only in mechanical properties but also in chemical composition. Some of this variability seems to be built into the genome, and some is the outcome of production procedures. The condition of the spider, both external and internal, affects both silk production and mechanics. Silk seems optimised for a wide range of conditions, and rapid temporal adaptation to the environment seems important. Adaptability is advantageous for silks used in orb webs that are often tested to the limits of their design criteria. Typically, such webs are rebuilt every day, and daily changes in climate (ambient temperature and humidity) or type of prey will affect the conditions under which the silk performs.

Silk, like many manmade fibers, is viscoelastic and has a very small plastic component that in most silks can be overcome by curing with water. The shapes of stress-stain curves vary greatly among different silks of the same spider as well as among the same type of silk in different spiders (Figure 3). This is due to differences in both spinning dope and spinning conditions.

The importance of spinning conditions cannot be overemphasised and is to a considerable degree under the spider's control. The animal, for example, can vary spinning conditions by (1) adjusting its running speed when laying down silk, (2) modifying the speed in which it pulls silk with its leg, or (3) building its webs during different times of the day or night—which affects not only spinning temperature but also spinning speed in this exothermic creature.

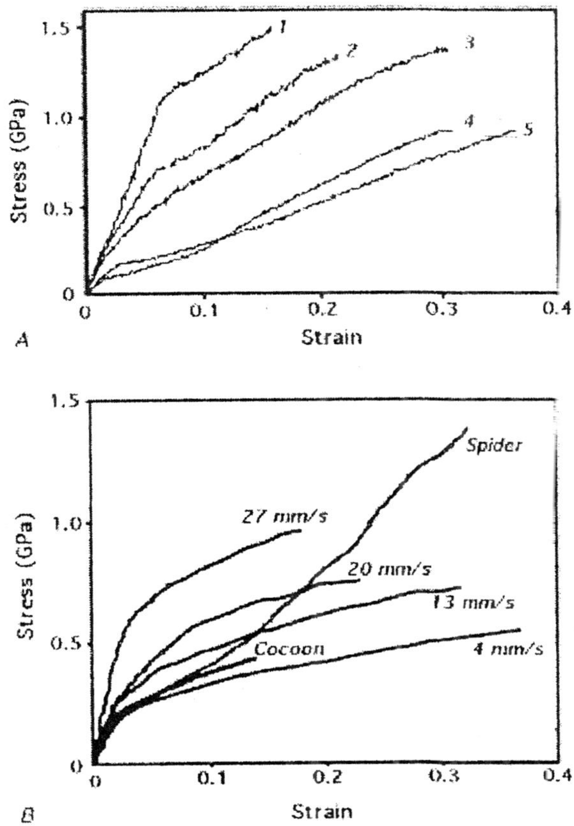

Figure 4. The mechanical properties of spider and insect silks (a) Spiders: Stress-strain characteristics of silk reeled from spiders belonging to widely diverging taxa: (1) Euprosthenops sp.(Pisauridae), (2) Cyrtophora citricola (Araneidae), (3) Latrodectus mactans (Theridiidae), (4) Araneus diadematus (Araneidae), and (5) Nephila edulis (Tetragnathidae). (b) Bombyx mori: Comparison of spider silk and manmade fibers with Bombyx silks drawn at different speeds.

Not only spider silk is affected by spinning conditions. In the silkworm Bombyx mori (and presumably also other lepidopteran silks), spinning determines to a large extent the mechanical properties of the silk.

It is generally assumed that commercial silkworm silk is much weaker and less extensible than spider dragline silk. However , silkworm silks reeled under specific conditions can approach spider dragline silk. Commercial Bombyx mori silkworm cocoon silk has a tensile strength of about 0.5 GPa, with a breaking elongation of 15% and a breaking energy (toughness) of 6´10^4J kg^{-1}.

Typical Nephila spider dragline silk has a strength of 1.3 GPa, with an elongation of 40% and a toughness of 16×10^4J kg^{-1}. In both silks, these mechanical measures show considerable inter- and even intraindividual variability. While spider silk for analysis is reeled evenly from the immobilised animal, silkworm silk typically is obtained from the cocoon. This means that it is spun by the mobile silkworm accelerating and decelerating its head in a figure-eight arc with attachment points at each reversal of direction. However, artificial reeling from immobilised silkworms under steady and controlled conditions can produce fibers that are clearly superior to the naturally spun controls (Figure 4b).

For example, at a silking speed of 4 mm s^{-1}, the breaking elongation of Bombyx silk is comparable to spider silk (37% versus 35%), and at a speed of 13 mm s^{-1}, the breaking energy of Bombyx silk approaches that of Nephila silk (12×10^4 J kg^{-1} versus 16×10^4 J kg^{-1}), although breaking strength does not (0.7 GPa versus 1.3 GPa). At even faster speeds, silk toughness decreased mainly because of loss of extensibility. It remains to be seen whether the residual differences between the silks of silkworm and spider are due to specific differences in the composition of the principal silk molecules or to their arrangement.

6

Testing of Raw Silk

Today, a number of varieties of silk fabrics are produced on handlooms and sophisticated power looms. This requires different qualities of raw silk. In order to assist the weaving industry in the selection of the required raw silk, it must be first tested and classified. Further, the raw silk reeling industry requires well-defined standards, which can only be achieved by silk testing. As the demand for silk is global and a number of countries compete in the trade of raw silk, it is necessary that there should be industry standards for raw silk quality so as to enable buyers to purchase raw silk at internationally accepted grades. This is the reason why all raw silk produced should be classified following testing.

The testing of raw silk is based on the procedure laid down by the International Silk Association (I.S.A.). This procedure is quite conservative and the equipment used is consistent with traditional patterns when compared to procedures followed for general textile products. The traditional method has been widely preferred in silk producing and consuming countries of the world. The mechanical testing procedure is similar everywhere, but the compilation of test results and standards for various grades differs slightly from country to country.

In 1928 and 1929, International Technical Conferences were held in Yokohama and New York, respectively. All silk producing countries such as Japan, China, Russia, France, Italy, etc., and silk consuming countries such as the United Kingdom and the United States, were represented. The organisation of the International Silk Association and the organisation of the third International Technical Raw Silk Conference in Zurich in 1949 followed by a fourth one in New York in 1950, contributed to placing silk technology and silk research on a global basis.

In 1949-1950, a Bulletin was issued by the International Silk Association (I.S.A.) on International Standard Methods for Raw silk Testing and Classification. In 1961 a revision was made and it was decided to divide them into three categories. A second amendment was made in raw silk testing and classification in 1974. Recently, an electronic testing system was introduced into I.S.A.'s methods. Two main tests are usually applied, namely the conditioned weight test and the test for quality. The quantity tested from a consignment depends on the total weight. Usually, 5-10 bales of 60 kg each or 20 bales of 30 kg each is the unit of testing used for the export of raw silk.

Quality Test

Visual Inspection

This test is conducted on the whole lot of raw silk, which has to be classified. This is carried out in an inspection room, which is well illuminated. There are three main factors that have to be tested. These are:

Uniformity

In this test, the entire lot is inspected to assess the

uniformity of colour, lustre and feel. The result is not recorded in the I.S.A. method, but is classified as good, fair and inferior under the Japanese method.

General finish

Here, considering the presence and degree of a number of defects assesses the general finish of the lot. These defects are:

i) *Re-reeling*. Gummed skeins; gummed spots on skeins; double ends; irregular traverse and partial flack of traverse.

ii) *Finish*. Tangled filament, defective lacing, filament out of place in skein (pulled filaments loose).

iii) *Arrangement*. Lacing of booking card through skeins; non-uniform skeins; wrong twisting; raised filament; streaky filament; cut ends; discoloured skeins; foreign matter on skeins; irregular skeins on book; knots on skeins; skeins or books of different types.

iv) *Damage*. Books of irregular shape; gummed books, soiled filaments; frayed skeins; insect attached skeins; musty skeins, etc.

The results are expressed in terms such as good, fair, poor or inferior. Outstanding defects are mentioned in the Test Certificate.

Nature

The degrees of Colour and Lustre and the Smoothness or Hand of the lot are inspected and indicated in the following manner:

i) *Colour*: light, medium or deep

ii) *Lustre* : bright, medium or dull

iii) *Hand* : smooth, medium or rough

Sample Test

After completion of the test on the whole lot, the following tests have to be conducted. To conduct these tests, it is necessary to take out 50 skeins off the lot if the skeins weigh below 120 g. Alternatively, 25 skeins are taken if the skeins weight over 120 g. Having extracted the required number of skeins, the following test is carried out:

Winding test

Equipment required:

— Winding frame

— Wheels and bobbins

Number of sample skeins drawn out for this test should be 40 skeins out of 50 skeins for a lot which weighs approximately 70 g each and 20 skeins out of 25 skeins drawn out for each skein weighing approximately 140 g.

When winding is started only the top half of the sample skeins should be wound. The winding should be carried out at a predetermined speed for a specific duration. The number of breaks that occur should be counted and noted. When breaks occur, it is necessary to note the cause of each break and this should be recorded.

Table 1. Average speed and winding period for winding test

Size under test	*Preliminary winding*	*Average speed (metre/min.)*	*Winding period (minutes)*	
			70 g skeins	140 g skeins
12 denier or finer	10	110	60	120
13-18	10	140	60	120
19-33	10	165	60	120
34-69	5	165	30	60
70 or coarse	5	165	20	40

The winding speed and duration of winding has also to be adjusted according to the denier of the raw silk being tested. For reference purposes, the different time and speed and other requirements are shown in Table 1.

Size deviation test

Equipment required:

— Sizing reel (1:125 metres in circumference)

— Balance

— Denier scale

Only a fixed length from each skein is taken for the testing. The highest degree of deviation is noted and compared with the average size deviation of the conditioned weight. When testing raw silk of 33-denier or finer in size, 200 skeins of 450 metres each are taken out by picking out four skeins out of each of the 50 sampling skeins taken from the testing lot. In the case of 34-denier and over, 400 sizing skeins are extracted by picking eight skeins of 112.5 metres in length each from the 50 sample skeins (total length 45 000 metres).

Test

The entire lot of sizing skeins should be in ten separate lots. Each sizing skein should be weighed on a quadrant scale in each lot separately. The total weight of all the skeins in each lot is therefore obtained. Table 2 indicates the size deviation of the skeins being tested. This table covers the number of skeins in a lot, the graduation in the scale for weighing each skein and all the skeins of a lot together.

If there is a difference in weight beyond the permissible denier indicted in the below table when the weight of the individual skeins are totalled and compared with the result obtained by weighing all the skeins together, the weighing operations both of single skein and the whole lot must then be repeated.

Table 2. Number of skins in a lot and graduation in scale

Sizes	*No. of skeins in a group*	*Accuracy of scale For 1 skein*	*For group*	*Permissible range of denier*
330 denier or finer	20	0.5 denier	0.5 denier	1.5 denier
33-49 denier	40	1.0 denier	2.0 denier	4.0 denier
50-99 denier	40	2.0 denier	2.0 denier	8.0 denier
100 denier or coarser	40	5.0 denier	2.0 denier	19.0 denier

The number of individual sizing skeins of each denier size should be noted in the table of frequency distribution.

The size deviation can be obtained by the following formula:

$$S = \sqrt{\frac{\sum fi(ti - M)_2}{N}}$$

Where t_1, t_2, t_3 ...., tn are the sizes of the sizing skeins in a denier; f_1, f_2, f_3 ...., f_n are their frequencies.

N is the total number of sizing skeins.

M is the calculated average size.

The results of the size deviation test are calculated down to two decimal places.

— *Maximum deviation test*: the difference between average size and average of four thickest sizes and also the difference between average size and average of four thinnest sizes are determined.

Comparison is then made of both differences and the result of maximum deviation determined as the larger one. The result of maximum deviation is calculated down to one decimal place.

— *Average size test*: All the sizing skeins are then placed in the conditioning oven and dried at 140°C. Until it reaches a constant, weighing is undertaken after the first ten minutes of drying.

$$\text{Size deviation(denier)} = 0.5\sqrt{\frac{1}{100}(1706 - \frac{25281}{199})} = 1.90$$

Maximum size deviation (denier) =

$$\text{Ave. of 4 coarsest size} = \frac{24.5 + 24.0 + 24.0 + 23.5}{4} = 24.0$$

$$\text{Ave. of 4 finest size} = \frac{(16.0 + 16.0 + 16.5 + 16.5)}{4} = 16.25$$

Coarsest deviation = 24.0 - 20.205 (Ave.size) = 3.795

Finest deviation = 20.205 (Ave. size) - 16.25= 3.945

Therefore, the maximum size deviation is 3.945.

Evenness test

The test is carried out with test samples of a fixed length using a Seriplane. These test samples represent fine passages and coarse ones, divided into 3 groups (Evenness Variation, I, II and III) according to the degree and fequency of size variations.

— *Evenness Variation I:* The intensity of variation greater than the V0 panel but does not exceed V1 panel of the Standard Variation Photographs.

— *Evenness Variation II:* The intensity of variation greater than the V1 panel but does not exceed the V2 panel of the Standard Variation Photographs.

— *Evenness Variation III:* The intensity of variation, which includes all the variations greater than the V2 panel of the Standard Variation Photographs.

Panel—A panel is a section of raw silk 127 mm wide by 457 mm long uniformly wound from a bobbin on to an inspection board.

In the Japanese method, Variation I is applied not for the major test or auxiliary test but for the optional text.

— *Apparatus and equipment:* Seriplane (127 x 457 mm), Standard Photographs and Illumination room.

— *Sample:* The sample for the test consists of a total of one hundred panels from 50 test samples taken at the rate of two panels from each test sample.

The thread is spaced on the inspection panel according to the size under test as follows:

9 denier or finer	133	threads per	25.4 mm
10 to 12 denier	114	"	"
13 to 16 denier	100	"	"
17 to 26 denier	80	"	"
27 to 36 denier	66	"	"
37 to 48 denier	57	"	"
49 to 68 denier	50	"	"
69 to 104 denier	40	"	"
105 to 149 denier	33	"	"
150 to 197 denier	28	"	"
198 denier or coarser	25	"	"

Test

The test is conducted by the estimator from a position of about 2 metre distance directly in front of the inspection panels, which are placed in such a way that the panel receives the same intensity of light by indirect lighting, over the entire board. On any one side of the inspection board, each stripe found on each panel is carefully compared with the standard variation photographs and the intensity of variation is determined. The frequency of evenness Variation I, II and III found with the 100 panels is recorded separately. The record shows each total frequency of variations grouped in Evenness Variation I, II and III.

The evaluation of evenness in the Indian method is indicated by percentage. Percentage is indicated to the nearest 5 percent starting from 100 percent to 50 percent. Below 50 percent, it is made to the nearest 10 percent. The record then shows the estimated evenness percentage of each panel, the average evenness percentage of a total of one hundred panels and the low evenness percentage of low panels corresponding to one-quarter of the total panels inspected.

Cleanness test

This test is conducted to ascertain Super Major Defects, Major Defects and Minor Defects. Each defect carries penalty points and the difference of the total penalty points from 100 gives the test result.

Cleanness Defects: These are categorised into three general groups, viz., Super Major Defects, Major Defects and Minor Defects .

Super Major Defects: All major defects in length or size which are ten times larger than the minimum size of Major Defects are named Super Major Defects.

Major Defects: These are divided into five as follows:

i) *Waste.* This is a mass of tangled cocoon filaments or fibres attached to the yarn.

ii) *Large slugs.* These are somewhat thickened places in the thread 7 mm and above in length, or very badly thickened places shorter than 7 mm.

iii) *Bad casts.* These appear as abruptly thickened places in the yearn due to the cocoon filaments not properly adhering to the raw silk yearn, or caused by feeding more than one cocoon filament at a time.

iv) *Very long knots.* These are knots, which have loose ends, 10 mm and over, or those made by incorrect tying of threads.

v) Heavy corkscrews are places in which one or more cocoon filaments are longer than the rest, and give the appearance of a very coarse and large spiral.

Minor Defects: The minor defects are subdivided into four as follows:

a) Small slugs, which are considerably thickened places in the thread from 2 to less than 7 mm in length, or extremely thickened places less than 2 mm in length.

b) Long knots are knots, which have loose ends below 10 and more than 3 mm in length.

c) Corkscrews are places in which one or more cocoon filaments are longer than the remainder, and give the appearance of a thick spiral.

d) Long loops or loose ends are loops or split ends, 10 mm and above in length, when measured along the filament.

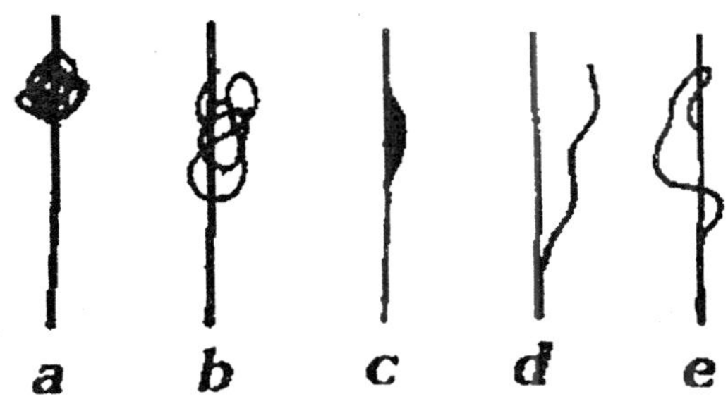

Figure 1. Various cleannessa)Waste, b) Slug, c) Bad casting, d) Split ends, e) Large loop

Apparatus and equipment: Standard Photograph for cleanness, Seriplane and lighting equipment.

Sample: The same as given under Evenness test.

Test

In this test the inspector stands in a position of about 0.5 metres (2 feet) directly in front of the inspection panels, under the same conditions of lighting as for the cleanness test. The actual number of cleanness defects of each class and kind of defect described above, are counted on the yarns on both sides of the inspection panel, omitting the parts on its edges. The class and kind to which each defect belongs is determined by comparing it with the Standard Photographs for cleanness defects. The record should indicate the number of defects in 100 panels found by testing and also show the cleanness percentage, which is determined by deducting from 100 percent the total penalty calculated by penalising each defect with the following rate:

For each super major defect .... 1.0 percent

For each major defect .... 0.4 percent

For each minor defect .... 0.1 percent

Neatness test

Neatness defects : Imperfection in raw silk yarn, which are smaller than those described as minor cleanness defects are known as neatness defects. Nibs are small thickened places or spots in the yarn less than 2 mm in length. Loops are small open places in the yarn caused by the excessive length of one or more cocoon filaments, less than 10 mm in length when measured along the filament. Hairiness and fuzziness are the conditions of yarn, which show small loose ends of less than 10 mm and fine particles of cocoon filaments protruding from the yarn. Small knots are knots, which have loose ends, less than 3 mm in length. Fine corkscrews are places in which one or more cocoon filaments are longer than the remainder and give the appearance of a spiral.

Apparatus and equipment: The Standard Photographs for neatness defects, Seriplane and lighting equipment.

Sample: The same as given in the Cleanness test.

Test

The test is conducted by the inspector, from a position of about 0.5 metres (2 feet) distance directly in front of the inspection panels, under the same conditions of lighting as for the neatness test. Each panel on any one side of the inspection board is carefully compared with the Standard Photographs for neatness defects and its neatness value is estimated in percentages. From 100 to 50 percent, the estimate should be to the nearest 5 percent. Below 50 percent, it should be made to the nearest 10 percent. The record should indicate the estimated neatness percentage of each panel, the average neatness percentage of a total of one hundred panels and the low neatness percentage represented by the average percentage of the low panels, i.e. on fifth (20 panels) of all panels examined.

Table 3. Deducting points by Neatness results

Neatness	*Deducting points*	*Neatness*	*Deducting points*
Above 80	0		
75	0.25	50	1.5
70	0.5	40	2.0
65	0.75	30	2.5
60	1.0	20	3.0
55	1.25	10	3.5

In the Japanese method, Cleanness and Neatness is represented as a mixed item by deducting the rate of Neatness results from the Cleanness percentage as shown in Table 3.

Tenacity and elongation test

To test the strength of the raw silk, the breaking point (g per denier) and the degree of elongation (percentage) is carried out on the Serigraph. This test is conducted in a room, which is maintained at a standard temperature of

Apparatus: Serigraph, sizing reel and scale.

Sample: Ten test pieces taken out of 50 test pieces. From each one of these ten test pieces, ten sizing skeins shall be prepared for tenacity and elongation test.

Test

The sizing skeins to be tested are placed in a room, where standard humidity can be maintained, for a sufficient amount of time to allow them to become adjusted to standard conditions.

Each sizing skein is then tested for tenacity and elongation using the serigraph, which is placed in the room under the same standards of humidity. The tenacity is expressed in grams per denier, while the elongation is expressed in percentage of total stretch of the portion tested. The result is indicated by the average results of ten sizing skeins. The result of tenacity is calculated by omitting the figures after the second decimal.

Cohesion test

By means of the Duplan cohesion tester, the number of frictions required to split silk thread for the purpose of examining the state of cocoon filaments sticking together, can be counted. This test is conducted in a room kept at standard temperature and humidity.

Apparatus: Duplan cohesion tester.

Sample: the sample for the test should consist of 20 test pieces taken out of 50 test pieces. The yarn should be free from any cleanness defect or apparent evenness defect in the portion, which is to be tested.

Test

The test is performed in a room where standard humidity can be maintained during the test. The maximum speed of stroke should be 140 strokes per min. The machine should be stopped after every ten strokes and every single yarn inspected very carefully to see if there are any open places. As soon as ten different open spaces, 6 mm long and above are observed, they are recorded against the number of strokes which should be considered as the strokes of the thread opened. The record of the test is the average number of strokes of 20 test pieces. The result of cohesion is recorded by omitting the decimal.

Quantitative Test

Conditioned weight test

When weighing the gross weight and the dried weight of raw silk, the conditioned weight is calculated by adding 11 percent (International Standard Regain) to the dried weight.

The conditioned weight of raw silk is obtained as follows:

— All skeins of raw silk for the test are put into a room with a constant temperature and humidity for 12 hours. 20 skeins are picked up for a test unit and are then weighed separately as the original weight.

— Two skeins are selected from the sample for the conditioned weight test.

— The sample skeins are dried in an oven one by one with the drying temperature at 140°C. The drying is continued until the different at every weighing (every 5 minutes) is within 0.1 g.

Moisture regain is calculated as follows:

— If the different between both skeins in moisture regain is over 0.5%, another one is tested again.

$$\text{Moisture regain (\%)} = \frac{(\text{original weight} - \text{dry weight})}{\text{dry weight (g)}} \times 100$$

The total conditioned weight of the unit could be obtained from the total original weight multiplying by the following coefficient.

The coefficient of the conditioned weight is obtained by:

$$\text{Coefficient t} = \frac{1.11}{1 + \text{moisture regain (\%)}}$$

Notice:

$$\text{Moisture content (M.C.)} = \frac{W - W'}{W} \times 100$$

$$\text{Moisture regain (M.R.)} = \frac{W - W'}{W} \times 100$$

Where, W is the original weight and W' is the dry weight.

Raw Silk Classification

Grades

For classification purposes, raw silk is divided into three categories according to their sizes:

1st category : 18 denier and below

2nd category : 19 to 33 denier

3rd category : 34 denier and above

The grades are expressed in the following order in all categories of raw silk:

4A, 3A, 2A, A, B, in I.S.A. Standard

and 5A, 4A, 3A, 2A, A, B, C, D in Japanese method

Grading with the major tests

The grade of a lot in sizes 33 denier and finer (categories I and II) are determined according to the lowest respective percentage of its Size Deviation, Evenness I, Evenness II, Cleanness, Average Neatness and Low Neatness. Whereas the grad eof a lot in sizes 34 denier and coarser (category III) are determined according to the lowest respective percentage of its Size Deviation,

Maximum Size Deviation, Evenness I, Evenness II, Cleanness, Average Neatness and Low Neatness as tabulated in the classifications table for each category.

Should any one or more of these results fall below the minimum limits prescribed for a grade, the lot is relegated to the lowest grade.

Degrading with the auxiliary tests

a) If any one of the Maximum Size Deviation, Evenness III, Winding, Tenacity, elongation or Cohesion tests of a lot in sizes 33 denier and finer, or if any one of the Evenness III, Winding, Tenacity or Elongation tests of a lot in size 34 denier or coarser, is found to be lower than the required value listed in the class of auxiliary tests, then the grade provisionally established in accordance with the preceding paragraph is lowered by as many grades as the difference that exists between the required auxiliary test class and the class actually found. Any difference that is more than one class is deemed as one class is deemed as one class difference with respect to the Maximum Deviation and Evenness Variation III of a lot in sizes 33 denier and finer and Evenness Variation III of a lot in sizes 34 denier and coarser. In case two or more auxiliary test classes are found to be lower than the corresponding values listed in the classes of the auxiliary tests, then the lot is declassed to the lowest auxiliary test class actually found.

b) In case the result of the visual inspection of a lot is found to be "Slightly Inferior" in its general finish and/or the result of the skein finish inspection in the Winding test of a lot is found to be "Poor", the grade of the lot is the one below that as has been determined in accordance with the preceding paragraphs.

c) In case the result of the visual inspection of a lot is found to be "Inferior" in its general finish, or in case the number of breaks in the Winding test exceed the limits mentioned below, the lot is determined as B grade (D grade in the Japanese method).

12 denier or finer.......................breaks 50(75)

13 to 18 denier ....breaks 40(65)

19 to 25 denier ....breaks (60)

19 to 33 denier ....breaks 35

26 to 33 denier ....breaks (55)

34 to 69 denier ....breaks 25(45)

70 denier or coarser ....breaks 20(35)

Miscellaneous Tests

Boil-off Test for Raw Silk

a) United States Method

— *Object*—The purpose of the Boil-off test on raw silk is to determine the percentage of sericin and water-soluble substance which the silk contains.

— *Apparatus* -

a) *Boil-off kettle* — A suitable receptacle for boiling off the silk.

b) *Oven* — Conditioning oven with forced ventilation efficient to dry the skeins within the time specified, positive value control, capable of drying the sample skeins at 140°C. The conditioning ovens should be equipped with a balance arranged to weigh the skeins with an accuracy of one centigram while suspended within the drying chamber, the hold of the skeins to be of such type as to insure free access of the dry air to all skins.

— *Water*—The water used for all parts of the test should be zero hardness.

— *Soap*—The soap should be properly saponified "neutral" soap in chip or

Chemical Requirements

— Moisture and volatile matter at 105°C, max. percent 10.0

— Sum of free alkali or free acid, total insoluble matter in alcohol, and sodium chloride, max. percent ...4.0

— Free acid calculated as oleic acid, max. percent 0.2

— Free alkali, calculated as NaOH, max. percent 0.2

— Insoluble matter in water, max. percent 1.0

— Titer of mixed fatty acids prepared from soap ....16 to 26°C

— Anhydrous soap content, min. percent 85.0

— *Samples*—Ten skeins, not more than one skein from one book, are taken to represent a five or ten bale lot. Approximately 10 g are removed from each skein. The silk removed from the original skeins is grouped into two parts of approximately 50 g each and marked, "part number one and part number two".

— *Dry weight before boiling*—Dry samples Part 1 and Part 2 separately in a conditioning oven at 140°C to constant weight as determined by successive weighings at five minute intervals. The first weighing is made at the expiration of the first fifteen (15) minutes of drying. The second weighing, made five minutes after the first weighing, is taken as the dry weight provided that the loss between the successive weighings does not exceed one centigram.

— *Boiling*—The samples are boiled for 45 minutes in a one percent soap solution. The weight of the soap should be 25 percent of the weight of the sample and the water 100 times the weight of the soap.

— *Rinsing* = Rinsing should take place in two baths containing water at 60°C, the volume of which should not be less than 25 times the weight of the sample.

— *Boiling—Second boil*—The samples are boiled again for 30 minutes in 2.5 litres of water per 100 grams of silk, i.e. 25 times the original weight of the sample (no soap).

— *Rinsing*—Repeat the rinsing operation.

— Centrifuge and dry at room temperature.

— *Dry weight after boiling*—Dry and weigh the samples using the same procedure as described above.

— *Loss in boiling-off*—The difference between the dry weight of the combined weight (Parts 1 and 2) before boiling and the dry weight of the combined weight (Parts 1 and 2) after boiling is the loss in boiling-off. The percentage of loss in boiling-off should be calculated individually per Parts 1 and 2. If the percentage of loss for each of the two parts differs by more than 1 percent, the test should be repeated. The average results of the first test and the repeat test should be reported.

b) European method

— Object—The object of the boil-off test on raw silk is to determine the percentage of sericin and water-soluble substances, which the silk contains.

— Apparatus -

a) *Boil-off kettle*—A suitable receptacle for boiling off the silk.

b) *Oven*—Conditioning even with forced ventilation, positive valve control, capable of drying the sample skeins at 140°C. The conditioning ovens should be equipped with a balance arranged to weigh the skeins with an accuracy of one centigram while suspended within the drying chamber, the holder of the skeins to be of such a type as to insure free access of the dry air to all skeins.

— *Soap*—The soap used should be a neutral soap made from pure olive oil or olive oil foots and soda.

— *Samples* -

a) Ten skeins, not more than one skein from any one book, are taken to represent a five or ten bale lot. Approximately 10 g is removed from each skein. The silk removed from the original skeins is grouped into two parts of approximately 50 g each and marked, part number 1 and part number 2.

b) In the case of a single bale, five skeins, not more than one skein per book, should be taken. Two samples of approximately 10 g each are removed from each skein. These ten samples are grouped into two parts, as in (a) above.

— *Dry weight before boiling*—Dry samples part 1 and part 2 separately in the conditioning oven at 140°C to a constant weight as determined by successive weighing at five minute intervals. The first weighing is made after 15 minutes. Weight is considered as constant when the difference between two consecutive weighings is less than or equal to 5 centigrams. The final weighing is taken as the dry weight.

— *First boil*—Boil parts 1 and 2 for 30 minutes in a soap solution composed as follows:

Weight of soap : 25 percent of net weight of silk.

Concentration : 8 g soap per litre of water which should be demineralised, or at least at 0° (hydrotimetric)

Volume of solution : 35 times the weight of silk.

Carbonate sodium (CO_3Na_2): 0.3 g per litre of water in the solution.

— *Rinsing*—Rinse by stirring and squeezing the silk by hand in demineralised water (or water at 0° Hydrotimetric). The temperature should be 60°C and the volume of water at least 25 times the weight of the silk.

— *Second boil*—Boil parts 1 and 2 a second time for 30 minutes in a solution having the same composition as the first one, except that no carbonate of soda is added.

— *Rinsing*—Rinse the silk in each of two baths, each time stirring and squeezing by hand.

— *First bath:* To consist of demineralised water (or water at 0° Hydrotimetric) at 60°C with addition of 2 or 3 g of CO_3Na_2 per litre. The volume of water must be at least five times the weight of the silk.

— *Second bath:* To consist of water at 60°C without carbonate of soda.

— *Drying*—Centrifuge and dry parts 1 and 2 for 2 hours at room temperature.

— *Dry weight after boiling*—Dry and weigh the samples using the same procedure as described in the above paragraph.

— *Loss in boiling-off*—The difference between the combined dry weights of parts 1 and 2 before and after the boilings expressed as a percentage of the combined dry weights as found in the above paragraph is the loss in boiling-off.

Exfoliation Test for Raw Silk

— *Definition*—Exfoliation in raw silk is the undesirable property of the individual filaments of silk split into very fine fibrils. These fibrils initially do not absorb dye as readily as the main fibres and show up after dyeing as fine white fuzz on the surface of the silk yarn and fabric. The Exfoliation Test, as outlined below, is made on raw silk and the degree to which the raw silk tends to exfoliate can thereby be determined.

— *Sample*—The sample for the Exfoliation Tests consists of twenty original skeins drawn at random, an equal number from each bale of the lot, from different parts of the bale. Not more than one skein from any one book except in cases where the lot is composed of fewer than twenty books, and then more than one skein may be drawn from one book as is necessary to make up the twenty skein sample. The sample skeins are wound on regular winding bobbins for a period of twenty minutes.

— *Preparation of panels*—Twenty panels, one from each sample bobbin, is reeled on the metal frames on a specially modified seriplane equipped for this purpose. The size of the panels is 4 and three-quarters inches high and 3 and one-quarter inches wide. The threads are spaced at 25 threads per inch. Each frame is made to hold five panels, making a total of four frames of five panels each for each test. The specially modified seriplane is so constructed that the silk may be reeled on either one or on two frames at the same time. After the silk is reeled on the frames, the frames are placed in special racks for degumming and dyeing. The degumming and dyeing racks are made to hold eight frames or two lots each. Either four or eight frames (one or two lots) may be degummed and dyed at one time.

— *Degumming and dyeing*—The degumming and dyeing are done at the same time in a solution made up as follows:

  — Approximately twenty-four litres of water are put into the degumming and dyeing tank and the water is heated. Sixty g of sodium metasilicate are thoroughly dissolved in one litre of hot water and poured into the tank. Thirty g of dye are thoroughly dissolved in another litre of hot water and poured into the tank. The solution in the tank is then stirred thoroughly. A special dye known as United States Testing Company Sky blue S due is used.

The thermometer should be left suspended in the solution at all times. When the temperature of the solution has risen to 195°-200°F, the racks holding the frames are placed in the solution and left there for 20 minutes. During the 20minute period, the temperature of the solution should be maintained between 195°-200°F. The same solution may be used a second time for one or two additional lots (four or eight frames). If the same bath is used a second time, tem g of due are added to the solution, but no additional sodium metasilicate is required. The added dye must be dissolved thoroughly in one-half (½) litre of water before being poured into the solution. Sufficient hot water should be added to the solution to maintain the level of the solution high enough to cover the frames. The same solution should not be used for more than two degumming and dyeing operations. It is necessary that the tank be drained and thoroughly rinsed at the end of every day even though the solution has been used for only one degumming and dyeing operation. After 20 minutes, the rack is removed from the degumming and dyeing solution and the frames are rinsed in the rinsing tank to remove excess dye. The rinsing tank is filled with warm water and rinsing is carried out by dipping the rack containing the frames up and down in the warm water three or four times. The silk on the frames is then dried thoroughly. The silk can be dried at room temperature or exposing the silk on the frames to a warm air blast or dried on the frames in a suitable warm oven. If the silk is dried in an oven, care must be taken not to scorch the silk.

— *Preparation of the panels for inspection*—It will be found that the silk threads on the frames will have a tendency to cling together in groups of three to six threads after the degumming and dyeing operation. After the silk has dried, the threads must be carefully separated before the silk is inspected. The separation of the threads can best be done while the frames are on the inspection racks. The threads can usually be separated by running a smooth glass rod, or similar instrument, horizontally across the threads near the top and bottom of the frame. In some cases, it may be necessary to separate some of the threads by means of a needle run vertically between the threads. These operations must be done on both the front and back surfaces of the frame. Extreme care must be taken in separating the threads in order to prevent breaking the filaments or threads. A limited amount of experience and practice will show the best way to separate the threads.

— *Inspection and grading of the panels*—The inspection and grading of the panels is done on the American Standard Inspection Rack. The two special holders are suspended over the top of the overhead light reflector so that the exfoliation frames can be placed in front of the light reflector, the top of the frame being approximately 1 and five-eighths inches below the forward edge of the reflector. A black seriplane board is placed in the upper brackets of the seriplane inspection rack behind the exfoliation frames. The seriplane board is inclined towards the inspection frames to an angle of approximately 40° from the vertical. The angle may be varied slightly so that the silk thread defects can be more easily observed. The exfoliation frames are maintained in a vertical position. Each of the twenty panels is given a rating by comparison with the standard photographs rated as follows: 90-80-70-60-50-30. Each panel is compared with the standard photographs and is given one of the following ratings in accordance with the standard: 100-95-90-85-80-75-70-65-60-50-40-30-20-10. If any filaments have been broken in the separation of the threads which have clung together after degumming and dyeing, care should be taken that the broken filaments are not mistaken for exfoliation and penalised. This difference can be detected after some experience in inspecting panels.

— *Calculation and rating reports*—The general average is obtained by calculating the average of the twenty individual ratings of the twenty panels. The penalty average is obtained by calculating the average of the lowest five panels.

— The classification degree is the average of the general average and the penalty average. The rating is based on the classification degree as follows:

Perfect	95 and higher
Excellent	85 to 94.99
Good	75 to 84.99
Fair	65 to 74.99
Poor	50 to 64.99
Very poor	10 to 49.99

Dye for exfoliation test

United States Testing Co., "Sky Blue S. Dye". This dye has the following formula:

	Class	*Colour Index*
Alphazurine A : 16 percent	acid	714(NAC)
Wool Violet 4BN : 15 percent	acid	698(NAC)
Glauber Salt : 69 percent		

The two (2) dyes (Alphazurine A and Wool Violet 4BN) are strong colours. For this Exfoliation test, Glauber salt could be omitted and less dye required to obtain the desired colour for the test. In that case, the following proportions are suggested:

	Class	*Colour Index*
Wool Violet 4BN : 48 percent	acid	698(NAC)
Alphazurine A : 52 percent	acid	714(NAC)

7

Health Care Measures in Food Plants

Muga food plants thrive well in sub-tropical areas. The main food plants are Som (Machilus bombycina) and Saulu (Litsaea polyantha) though some other food plants like Mejankari (Litsaea citrala), Digloti (Lisaea salicifolia), choa (magnolia sphenocarpa), Bhumloti (Symploces gradifflora), etc., are also there. Some plants grow up to an altitude of about 600 metres while Sualu up to 1000 metres above sea level. Suitable Som and Sualu plants are selected taking into consideration their growth, rearing results, disease resistance, etc. The leaves of som plants have different taste. They are to be selected considering of their growth, leaf-yield, shape and size of the leaf, its taste and odour. Naharpatiya, and Sompatiya with sweet taste are said to be the best.

Selection of Som plants should be done in Government Farms on the merit of their performance. There is only one type of Sualu plant. However three varieties of it may be differentiated according to the size shape and colour of leaves. Four varieties of Mejankari found according to the size, shape and colour of their leaves. Digloti is also found in four varieties-two from size and two according colour of its leaves.

All the muga food plants are perennial and as such the cost in maintenance of the plantation is less. Som trees are preferred by rearers due to their long life span and more resistance to stem borer's attack. Plants growth is controlled be several environmental factors like solar radiation, temperature, humidity, rainfall, wind etc. One should have the basic knowledge of environment, climatic zones, distribution of various plants, and ecology and plant physiology for a better plantation.

Disease and Pests of Muga Food Plants

Muga food plants are also attacked by various diseases and pests. They mainly depend upon the following parameters:

- Variety of the plants,
- Climate of the place,
- Ecological condition and
- Soil.

It is the primary duty of the planter to select the disease resistant suitable variety and good soil before plantation. Proper care must be taken to -protect the plants from diseases and pests including their enemies. In case of Muga food plants, prevention of the disease or pest is more important than control, as leaves are used for rearing of Muga worms. Common diseases of Muga food plants are:

- Leaf spot-caused by a fungus and it may be controlled by spraying insecticides with proper care.
- Grey blight- It is also caused by a fungus. To control this disease picking and burning of diseased leaves in early, stage are required.

— Red rust- caused by a parasite (alga). Pruning, removal of affected twigs and burning them is necessary to control this disease. Besides the above diseases some other diseases are also there for which controlling measure are to be taken in time.

There are many pests which damage the Muga food plants. Some of them are:

— Stem borer: It is also called carpenter worm. This may be controlled by applying fumigants soaking in cotton wool and putting it into the hole and close it with mud.

— Leaf gall: Gall is tumour like growth caused by different gall wasps. This may be controlled by plucking the gall infected leaves and pruning the trees.

— Sucking pests: Thrips, Jassids are the common sucking pests. These can be controlled by application of insecticides.

— Leaf rolls: It cause by insects secreting gummy substance by larvae forming roll of leaves. Leaf rolls are to be collected and burnt for control of this insect.

— Hairy caterpillars: Remarkable damage done by many hairy caterpillars like cricula. They are to be collected and burnt for controlling them.

Plants may have their deficiencies also. Various cultural practices will help in preventing/controlling many of the diseases and pests. Besides, mechanical measure is to be taken to kill the adult pests. Collection of infected leaves and live cocoons of various insects is to be made and burnt them. Closure of holes made by borers is also essential by required materials. Application of some chemicals may suitably be made after proper experimentation. Spraying of insecticides is to be done before 1-1½ months of rearing to avoid residual effect.

Diseases of Muga Worms

There are four common diseases of muga silkworm. They are pebrine, grasserie, flacherie and muscardine. Transmission of all the diseases except pebrine is mainly infectious. pebrine is infectious and inherited. The incidence of different diseases varies from place to place and season to season. Grasserie and Muscardine are very rare in case of muga.

Pebrine (Phootukia)

Pebrine is the most dangerous disease of muga silkworms. It is cause by protozoa (Nosema species). The spores of Nosema are bright and oval in shape when seen under microscope. It effects in all seasons but higher in autumn and winter seasons. The germ of pebrine loses its viability after six months in case of muga worms. It attacks in all stages of the life cycle of muga worm. Flies can carry pebrine germs infecting all worms in an outdoor rearing.

The important symptoms of this disease in larval stage are less appetite, worms becomes weak, sluggish, lose their natural luster and become dull in colour, unnatural small dots-like black spots appear all over the body especially in $4_{th}/5_{th}$ instar, and they become irregular in growth and moulting. Cocoons formed by infected worms are not uniform. Emergence of moths is very irregular. Infected female moths lay few eggs. The embryo dies inside the egg. The following precautionary measures should be taken to control pebine disease:

— Exposing the cocoons to high temperature (34° C) at pre-pupal stage for 16 hours a day with 55-65 %

humidity reduces the chance of pebrine infection of the eggs.

- Keeping the eggs in hot water at 40° C for 30 minutes after 36-40 hours of oviposition will also help in disinfection and in killing pebrine spores.
- Seed cocoons should exclusively be procured from the uninfected areas.
- Transportation of cocoons from infected to uninfected area should be avoided.
- Resistant stocks are to be encouraged for rearing.
- Infected cocoons should be used for reeling or spinning.
- Male moths, dead larvae, excreta from infected progenies and the smears of female moths should be burnt in pits at a distance and buried.
- Layings must be selected on the basis of strict microscopic examination and proper disinfection with formaline.
- If the infection is detected in a lot, all the worms along with plucked leaves of the particular tree should be burnt and buried.
- Microscopic examination in all stages of Muga is essential. The hands of persons are to be washed with per cent formaline to avoid contamination.
- The field staff should undertake survey works in the field from time to time so as to forecast the possibility of spread of disease.

Grasserie (Phularog)

It is a virus and contagious disease. It occurs in the gut and later to skin, fat bodies, blood cells, nerve ganglia, trachea, etc. It occurs mainly in summer and may occur in winter also. It is of rare occurrence in Muga. Worms lose appetite, discharge soft excreta become sluggish and immobile. The dody becomes soft, flabby and turns brownish, loses its natural shape. The worms die after about 24 hours of attack. It gets ruptured after death with a slight injury and omits obnoxious smell. Body colour turns to blackish and ultimately tissues are melted. Cocoons formed by infected worms are flimsy.

Worms die during spinning, after pupation and even on emergence/ Dead pupae melt and make the cocoons dirty. The infected moths possess flaccid body with crumpled small wings. On examination under microscope, polyhydral bodies are seen. They are nuclear and cytoplasmic in nature. It may be controlled by washing the eggs by formaldehyde, Sodium hydroxide, hydrochloric acid and Potassium permanganate of appropriate strength, following Sericultural hygience, selection of disease resistant strains, and suitable plantation for rearing, etc. It can be avoided by reducing summer rearing (mid June to August).

Flacherie (Mukhlage)

Flacherie is very common in Muga worms. It is caused by different kinds of bacteria. It is an infectious diseases. Some authors consider it as a virul disease. This disease infects the alimentary tract of the worms but blood cells, hypodermis and fat bodies may also be infected. Sudden fluctuation of weather is the main cause of the disease. Unsuitable leaves, overcrowding, heavy rain or long drought, etc., are also responsible for spread of the disease. Last stage worms are more susceptible to it. Worms die before spinning. It is common in rainy season.

Common symptoms of this disease are larva has swelling body, lethargic, motionless, weak and reduced feeding, body becomes soft, hangs its head downward and dies. They defecate coloured, sticky, semisolid excreta, chain of faecal beads; protusion of the recrum has been observed. Proper incubation of eggs, care of worms at theearly stages with quality leaves, protection from direct sun and rain, use of healthy and examined seeds, collection of seed cocoons from healthy and uninfected areas the precautionary measures to be taken for the disease. Worms should be handled less. Appropriated rearing season should be derermined carefully.

Mascardine (Bhekur Rog)

It is a fun fungal disease. The infected worms become harder, paler and completely inactive followed by bending of the body dorsally. It dies in about 20-24 hours of attack. Dead worms are compressed, reduced body fluid and spongy with fragile skin. A white encrustation appears round the body becomes laterally compressed, dry, hard, brittle and,mummidied. The infected pupae become mummified and hard like that of the larvae. It is of very rare occurrence in Muga. Its occurrence is reported during winter rearing only. This disease may be avoided by reducing the rearing of winter season (December to February).

Pests of Muga Worms

Common pests which damage the larvae of Muga are described below. The prevalence of these pests usually synchronises with rearing period of Muga worms.

Fly pest

Uzi fly is a parasite to larvae and pupa of Muga.The female lays eggs (30-40 Nos) on the body of the larvae. The eggs hatch out within three days. Maggots come out within 10-15 days. They pupate inside ground or in any dark corner and sometimes in the cocoons. They emerge out as fly ultimately. During winter season the pupae remain under hibernation. The adult will emerge in the following spring. Common fly (Trycolyga bombycis) attack Muga but the extent of damage is not so much. Sturmia sericariae attacks the Muga worms in the same way like other flies. The incidence of this parasite is less (about 15%) in spring and autumn crops but about 50% in winter and the late Summer Crops.

Ichneumon Wasp

The adult is 2-2.5 cm in length with two pairs of transparent wing and a pair of prominent antennae. The insect is yellow/black in colour having dorsolateral black spots on each sternum. The prothroracic shield also bears four black spots The female has nearly on centimeter long prominent needle like ovipositor with long stylets.al body by inserting its ovipositor ell.

The maggot after hatching consumes the entire pupa/ larva. It forms cocoons around the host, pupate and finally comes out as fly. It is found more during late autumn crop. Pimpla punctuator also infects Muga worms in the same way as the Ichneumon wasp and other species of the moths. Besides these, predators like

— Canthesone bug

— Red ants,

— Preving mantis

— Reduviid bag,

— Wasps, etc, are also damaging to the cocoon crop.

Controling Measures

The controlling measures are only mechanical as we cannot apply any insecticides or pesticides due to harmful effects on Muga worms. So, constant watch and wards is the only remedy to this problem, some attractants and chemesterilants are employed with success. The use of odour baited lure has been helpful in trapping various flies, wasps and predators. Tugon bait with fish refuse or flesh and Tugon bait with refuse of country liquor ar used and found effective. Thick liquid gum putting at the end of bamboo poles may be used to catch various flies, wasps, etc on the tree.

Lali of molasses mixing with Tugon bait may also be used to attract them. Pitcher plants may be planted around the site of rearing, which will also help us to kill flies, etc. Sex pheromone may be used for luring insects into lethal traps. It is found very useful in controlling harmful insects. It has been reported that males of saturniids can be lured to their females upto a distance of about 4 Km. Regular inspection of cocoonages (Jalis) is required to collect and destroy the fallen maggots on the floor. The floor of the cocooning room should be plastered so that maggot cannot go inside. Early stifling of reeling cocoons is to be done to kill the maggots of uzifly and others.

Besides disease and pests of Muga, there are many enemies, such as crows, bats, rats, ants, lizards, owls' kites, snakes, mice, monkey, jackal, etc. Careful watching is most essential to avoid damage of the crop. Birds are scared away by bows and pallets and by other bird scaring instruments. Rearers should be educated for controlling various pests, predators and diseases of Muga worms.

Mulberry Diseases

The leaf of mulberry (Morus spp) is the basic food of silkworm (Bombyx mori L). The role of leaf quality in the success of silkworm rearing and cocoon yield is well established. Mulberry is a fast growing plant and leaf contents are rich in protein, sugar and moisture. The nutritional value of leaf, however, varies greatly owing to a number of factors among which the diseases play an important role.

Being a perennial plant, mulberry is exposed to the environmental vagaries and pathogens prevailing in the area throughout the year. As a result, many fungi, bacteria and nematodes cause number of diseases. Feeding the leaves of infected plants to the silkworms adversely affects their growth and health resulting in partial or sometimes complete failure of cocoon crops and reduced marketing quality of cocoons produced. Hence, a close watch, timely control measures and regular monitoring are very essential to manage the problems posed by diseases.

Cercospora leaf spot

Causative organism: Fungus, Cercospora moricola

Initial symptoms of the disease are small light brown specks on leaves. These spots later enlarge into nearly circular, dark brown lesions with chlorotic margin. Whitish conidial masses appear on the lesions solitarily or in a circular fashion. Severely infected leaves turn yellow and fall-off prematurely. The loss due to the disease ranges from 10-20 per cent depending on the age of the garden, temperature and frequency of rain.

The disease occurs during South-West monsoon season. Maximum disease severity is seen during August to October. The disease may also prevail on plants during November and December depending on the receipt of North-East monsoon. Young leaves within the age of 18 days are susceptible to infection.

All the infected leaves should be collected and burnt prior to receipt of each rain to reduce availability of the inoculum in the garden for further spread. Spray of 0.2% Carbendazim (2 gram of chemical/litre of water) twice at 15 days interval is used as chemical control. Safety period of Carbendazim is 15 days.

Powdery Mildew

Causative organism: Fungus, Phyllactinia corylea.

White powdery patches appear on the under surface of leaves. Corresponding to these patches on the upper surface, the leaves turn chlorotic. Under favourable conditions, the symptoms may progress to form an almost total white superficial covering on the undersurface of leaves. As the lesions become older, they may turn black due to the growth of a hyperparasite on the mildew fungus. Infected leaves turn yellow, lose the moisture and become easily friable leading to defoliation. The loss incited by the disease ranges from 10-15 per cent.

The disease generally occurs during August to January. Maximum severity is noticed during October-December. Close spacing of plants and excessive nitrogen application favours disease severity.

Adopting wider spacing to facilitate adequate aeration penetration of sunlight helps to reduce disease severity. Removal of weak branches and weeds checks the humidity build-up under the plant canopy thereby reducing the severity of the disease. Chemical control of mildew becomes necessary in hilly areas and regions receiving heavy rainfall. Fungicidal spray of 0.2 per cent Carbendazim/Dinocap (2 gram/ml of chemical in one litre of water) should be applied as soon as the symptoms appear on plants which may be repeated after 15 days, if necessary. Leaves can be fed to worms after 15 days if Carbendazim is used and 20 days if Dinocap is sprayed.

Leaf Rust

Causative organism: Fungus, Peridiospora mori

Numerous, small, light brown specks appear on the under surface of leaves which later turn into dark brown rusty spots.

The disease is most common during winter months of November-January. Medium and coarse leaves are susceptible to infection. The loss due to this disease is about 5 per cent.

Timely utilisation of leaves during winter months is most essential to prevent inoculum build up in the garden. Chemical control using foliar sprays of 0.2 per cent Carbendazim/Chlorothalonil (2 grams of chemical/litre of water) can reduce the disease severity. Leaves can be fed to silkworms after 15 days.

Bacterial Leaf Spot

Causative organism: Bacterium, Pseudomonas mori

Lesions initially consist of small, irregular, water soaked areas on the under surface of leaves. Later, they enlarge and become dark brown to black necrotic lesions. In wet weather, a sticky bacterial ooze can be seen on the lesions. If young leaves are infected, they curl and drop off, while medium aged leaves turn yellow prior to defoliation. The pathogen can also cause lesions on tender stem portions. Yield loss as a result of disease is generally not high. Occasionally it may cause 1-2 per cent loss in yield.

The disease occurs in most mulberry growing areas during June-December.

Infected leaves should be removed and burnt to reduce the inoculum load in the garden. Chemical control using 100 ppm Streptocycline solution (1 gram of chemical in 10 litres of water) as foliar spray twice at 4 day intervals has been found to be effective. Residual toxicity to silkworm is absent.

Stem Canker

Causative organism: Fungus, Lasiodiplodia theobromae

Mulberry is propagated mainly through stem cuttings. The cut ends of the cuttings provide the main portal for the entry of many pathogens. One such important pathogen is Lasiodiplodia theobromae (Botryodiplodia theobromae), a wound parasite causing stem canker in mulberry. Canker is a disease causing death and /or decay of the bark. Disease symptoms in nursery are:

— Failure of cuttings to sprout.

— Sudden withering and death of sprouts.

— Discolouration and drying of stems and buds above the soil.

— Rotting and peeling off bark on stem below the soil surface.

— Black mycelial strands seen below infected bark.

— Black eruptions on the bark of the infected stem portion.

Disease symptoms on grown-up plants

— Greyish brown discolouration of the bark at the cut ends of the stem.

— Delayed sprouting.

— Death of buds and sprouts.

— Black eruptions on the bark in the infected region.

— Death of plants.

The above symptoms can be observed few days after the plants are pruned.

The pathogen is soil-borne and spreads through irrigation, farm activities, insects and wind. Effective management of stem canker can be achieved through cultural and chemical methods.

Cultural methods

— Roguing: This method involves removal and burning of the infected cuttings and saplings in nurseries. In established gardens, the affected stem parts should be cut and burnt.

— Field sanitation: Field sanitation is essential especially in the control of L. Theobromae which survives in the soil and on host debris. Destruction of crop debris by burning reduces inoculum build-up. Stalking of pruned stems and twigs by the side of mulberry garden should be avoided since it aids in the growth, multiplication and survival of the pathogen which serves as an important source of inoculum.

— Land use: Raising saplings repeatedly in the same site should be avoided as it helps in inoculum build-up in the soil, making it sick. Such sick soil supports a very high incidence of disease and mortality of cuttings.

— Planting season: Planting during winter months should be avoided since cold weather delays normal sprouting thereby predisposing the cuttings to infection.

— Proper pruning: Damaging the bark or buds while pruning should be avoided to prevent entry of pathogen. Pruning the stem close to the buds helps normal sprouting and prevents fast colonisation by the fungus. Cutting close to the buds should also be followed while preparing cuttings to avoid delayed sprouting.

Chemical control

— Pre-treatment of cuttings: The cuttings should be dipped in 0.2 per cent Carbendazim solution (2 grams of chemical in one litre of water) for 30 minutes prior to planting followed by a spray after 15 days planting.

— Protecting established plants: Soon after pruning, the cut surfaces of the stems should be dressed with a spray/smear of 0.2 per cent Captan or Carbendazim solution. This is a very important step to be followed in gardens where shoot harvesting is practiced for rearing silkworms. The treatment forms a toxic barrier on the wound surface preventing the entry of pathogenic inoculum.

— Soil disinfection: This method is suitable for the treatment of sick soils in nursery beds. Carbendazim or Captan at the rate of 8 kg/acre should be broadcast after loosening the soil followed by irrigation. After 10-12 days, well decomposed farmyard manure should be applied and thoroughly mixed with soil for planting the cuttings. Post treatment application of farmyard manure is a prerequisite prior to planting in treated soils.

Biological control

Nursery Guard, a bioformulation of Trichoderma pseudokoningii can be used for long term protection, better survivability and growth of saplings. Mix 1 kg of nursery Guard with 60 kg of finely powdered, well decomposed farmyard manure/neem cake. Moisten the mixture to a maximum 30% moisture, store it under shade in the form of a heap and cover with a wet gunny cloth. This enhances the multiplication of biocontrol agent in the mixture. Broadcast the mixture @ 2 kg/m over the well prepared nursery beds and thoroughly mix with soil by light digging. Apply the mixture only to the planting area in nursery beds.

Root Rot Disease

Causative organism: Fungus, Fusarium solani and F. Oxysporum

Fusarium solani is more virulent and cause maximum damage. The disease appears in all types of soil and climate throughout the year. The disease incidence with plant mortality ranges from 2-10%.

Sudden wilting/withering and defoliation of leaves followed by drying/death affected plants. The underground symptoms include decaying of root bark and rotting of primary and secondary roots. Rotten roots turn black and contain black powdery debris below the bark. The severely affected plants loose the hold in the soil. The disease is soil-borne and appears in isolated patches in the gardens and quickly spreads to surrounding areas through irrigation and farm operation.

Chemical control involves the root dipping of saplings in 0.2% Carbendazim solution for half an hour and planting in pits dusted with 5g Mancozeb. Some method of applications are briefly described below:

For existing plantation

As soon as the initial symptoms like wilting of leaves and drying of branches are noticed, dust 5g Mancozeb/plant after removing the soil around the infected plants to a depth of 5-6". And also apply the same dose of Mancozeb to the plants in 2-3 surrounding rows around the infected plants.

For replantation

— Uproot the severely affected plants and destroy. Prepare one foot deep pit and remove the soil including rotten stump of roots and allow the pit for sun drying for a month before planting. Prior to planting, dip the roots of the saplings in 0.2% Carbendazim solution (2g of chemical in one litre of water).

— Dust 5g Mancozeb/pit and plant the treated saplings in the pit and cover with fresh soil.

— Also apply the Mancozeb alone @ 5g/plant for 2-3 rows of plants surrounding the infected plants.

— Apply second dose of Mancozeb @ 5g/plant a month after replantation.

Biocontrol

A biofungicide 'Raksha', containing Trichoderma harzianum has proved to be effective and is under large-scale field trial. The biofungicide can be used for the management of the disease.

Method of application

Preparation of mixture of 'Raksha'

— Mix one kg 'Raksha' with 50 kg FYM/neem cake (sufficient for 1000 plants) and keep under shade for a week covered with a wet gunny cloth to enhance the multiplication of Trichoderma.

For existing plantation: As soon as the initial disease symptoms appear, apply the mixture of 'Raksha' @ 500g/ plant at a depth of 5-6" below the soil including surrounding 2-3 rows of plants. For new plantation, apply Raksha @ 500g/pit and plant the Mancozeb (0.2%) treated saplings in the pits (Mancozeb is not toxic to biocontrol agent). Apply second dose of Raksha after 3 months of replantation.

***Lasiodiplodia* Root Rot Disease**

Causative organism: Lasiodiplodia theobromae (Botryodiplodia theobromae)

Symptoms are:

— Sudden wilting and withering of leaves.

— Drying and death of branches and plant.

— Appearance of black eruptions (pycnidial clusters) on the bark near the soil surface.

— Rotting and pealing of bark from roots.

— Greyish black mycelial strands appear on the under surface of bark and on wood portion of root.

— In presence of adequate soil moisture, rotting and pealing of bark extends even to the basal portion of the stump.

The disease occurs throughout the year. This disease can be controlled by:

— Pruned stem portion should not be stacked near the garden to prevent multiplication of the pathogen and further spread.

— Uproot and burn the diseased plant along with roots.

— Soil from where the infected plant was uprooted should be thoroughly drenched with 0.2% Carbendazim or Captan solution. After 15 days, apply well decomposed farmyard manure and plant a new sapling with its root pre-treated with 0.2% Carbendazim solution.

— Immediately after pruning, the cut surfaces and the stumps should be sprayed with 0.2% Carbendazim.

Root Knot Disease

Root knot is one of the major diseases limiting crop production throughout the world. It occurs throughout the year mainly in sandy soils low in organic matter. The severity of the disease increases with increased age of the garden. The estimated yield loss due to the disease is 15-30 per cent. Infected plants become weak and predisposed to other disease while severely infected plants ultimately die.

Causative Organism

The organism causing root knot disease, is a nematode (Meloidogyne incognita), an endoparasite inhabiting mulberry roots. Symptoms are:

— Stunted growth.

— Poor and delayed sprouting.

— Reduced leaf size and yield.

— Chlorosis and marginal necrosis of leaves, yellowing and wilting of leaves inspite of adequate soil moisture availability.

— Death of plants in severe cases.

These symptoms first appear in isolated patches, slowly spreading over the entire garden. Symptoms on the underground parts are:

— Formation of galls/knots on roots.

— Reduced and stubby root system.

— Retarded root growth.

— Necrotic lesions on the root surfaces and death of active rootlets.

This disease can be controlled by different methods.

Cultural methods

— Saplings free of nematode infestation should be used for planting to prevent the introduction of nematodes from diseased area into a disease free area.

— Heavily infested soils should be deep ploughed to expose the nematodes and their eggs to solar radiation. Keeping the land free of vegetation during summer kills the nematodes due to desiccation by solar radiation.

— Application of neem cake @ 2 MT/ha/year in 4 split doses reduces the disease incidence.

Chemical methods

Chemical methods of nematode management become necessary in sick soils and heavily infested gardens as they give quick results. Furadon 3 G (40 kg/ha) can be applied either in furrows of broadcast to the soil after light harrowing followed by irrigation. The leaves from treated plots can be fed to silkworms after 45 days.

Biological control

A bionematicide verticillium chlamy- dosporium can be applied with neem cake. One kg bionematicide to be mixed with 200 kg FYM and 24 kg neem cake is sufficient for 1000 plants. Prior to application, the mixture should be kept in shade for a week covered by wet gunny cloth. Apply the mixture @ 200 g/plant around the root zone and repeat the treatment every 4 months for 2 years.

Mulberry Pests

Mulberry Leaf-Roller, Diaphania Pulverulentalis

The leaf-roller, Diaphania pulverulentalis (Hampson) (Lepidoptera: Pyralidae) is a pest of mulberry. The leaf-roller infestation causes considerable reduction in leaf yield, resulting in economic loss to sericulturists. The infestation is observed in mulberry plantations from 15 days through 70 days after pruning or leaf harvest. Up to 100% infestation is recorded in several areas. Though, leaf-roller infestation is observed round the year but for May, infestation is severe from July to November.

Symptoms

The young caterpillars within the unopened young leaves secrete delicate, white, silky filaments, which bind the leaf blades together and feed on the soft green tissues on the leaf surface. Sometimes a single leaf is rolled up and the caterpillar is found within this rolled up leaf formed by the filaments and hence the name leaf-roller. Grown up caterpillars fed voraciously on tender leaves and their faecal matter can be seen on the leaves below the affected portion.

The leaf-roller not only feeds on the leaves at the apical portion of the affected plants, but also damages the young shoot. As a result, the plants become stunted. In such plants, tender leaves required for chawki rearing become scarce. In addition, the leaf-roller is highly susceptible to pebrine disease of silkworms. The pest larva infected with pebrine spores in the field can contaminate mulberry leaves. When such contaminated leaves are fed to silkworms, it can cause an outbreak of pebrine disease in silkworms. Thus, this pest causes dual damage.

Life-cycle

Moths are greyish brown in colour and measure about 1 cm in length. Wings have brownish wavy stripes. Moth lays 1 or 2 eggs on the apical portion of each shoot. Each female lays about 100 eggs. Incubation period is 3-4 days. Grown up caterpillars are greenish brown in colour with black markings on the lateral and dorsal regions of the body segments. The head is black in colour. Larval duration lasts for 10-15 days. Pupae are dark brown in colour. Pupation occurs in dry leaves and pupal duration is 8-10 days. The pest completes its life-cycle in 21-29 days.

Controlling methods

Adopt an integrated pest management (IPM) strategy comprising cultural/mechanical, chemical and biological control methods.

Cultural/Mechanical measures

— Clip the affected apical portions, collect them in polythene containers and dip in 0.5% soap solution (5g of soap in 1 litre of water) to destroy the larvae.

— Collect and burn dry leaves and weeds found on the surface of soil in infested mulberry gardens to destroy pupae.

— Dress mulberry shoots before transporting them to other areas for new plantations in order to check the pest from spreading.

— Promote the activity of bird predators by providing them with water in small containers in and around mulberry plantations.

Chemical Control: Spray 0.076% DDVP [1 ml of DDVP (76% EC) in 1 litre of water] on the apical portion of plants when infestation is noticed. A second spray is necessary 10 days after the first spray, if infestation exceeds 20%. Use mulberry leaves for silkworm rearing 10 days after spray. About 150 litres of DDVP solution is required/spray/acre. Freshly prepared solution should be used. The aforesaid management practices should be adopted simultaneously by Sericulturist in pest infested areas.

Precaution

- Undertake the insecticide spraying during the cooler hours of the day.
- Check the expiry date of insecticides while purchasing.
- Keep the container of the insecticides airtight after use.
- Keep the insecticides away from reach of children and pet animals.

Bihar Hairy Caterpillar, Spilosoma oblique

The Bihar hairy caterpillar, Spilosoma obliqua Walker (Lepidoptera: Arctidae) is a polyphagous pest infesting several crops including mulberry, sunflower, green gram, soya, jute, etc. It occurs almost throughout the year and infestation is severe from August to December and also January.

Symptoms

Young caterpillars are gregarious and they feed on the green soft tissues of tender leaves, leaving behind only the veins. Grown up caterpillars are solitary and feed voraciously on entire mulberry leaves, thus defoliating the plants.

Life-cycle

Moths have light brown coloured wings, brick red abdomen, peppered with dark row of spots laterally and dorsally. Gravid females lay 1,000 to 2,000 eggs in a group on the under surface of mulberry leaves. Eggs are pale green in colour and hatch in 5-7 days. Caterpillars moult 6 times. Grown up caterpillars measure 4-5 cm in length and are covered with bristles. Anterior and posterior ends of the caterpillar are black in colour, while the rest of the body is reddish brown.

There are seven instars and the larval duration is 27-32 days. They pupate in crevices found between stones on bunds, crevices of tree trunks and corners of nearby buildings. Pupae are dark brown in colour and measure about 2 cm in length. Pupal duration is 12-14 days. The life-cycle is completed in 44 to 53 days.

Management Practices

Adopt an Integrated Pest Management (IPM) strategy comprising culturally/mechanical, chemical and biological control methods.

Cultural/Mechanical measures:

— Collect and destroy egg masses and gregarious young caterpillars by crushing them.

— Use light traps to attract moths by placing a tray with 0.5% soap solution below the light source.

Chemical Control: Spray 0.15% DDVP [2 ml of DDVP (EC=76%) in 1 litre of water], 15 days after release of T. chilonis. Mulberry leaves ca be fed to silkworms 10 days after spraying.

Biological Control: Trichogramma chilonis is an egg parasitoid of many lepidopteran pests. It is widely used as a biocontrol agent of several crop pests. Release twice

T. Chilonis at the rate of 5 tricho-cards (20,000 parasitised eggs in each tricho-card) per acre, at an interval of 3 days. Parasitoid releases have to be undertaken 20 days after pruning or harvesting.

Cutworm, Spodoptera litura

The cutworm, Spodoptera litura F. (Lepidoptera: Noctuidae), is a polyphagous pest infesting several crops, such as mulberry, tobacco, castor, several vegetable crops, etc. It occurs sporadically on mulberry, where vegetable crops are commonly grown. It occurs mostly in between August to December and upto February months.

Symptoms

The cutworm caterpillars attack shoots of young mulberry plants and cut them, hence the name cutworm. The cut portion of the shoot dries up and falls off. They also feed on mulberry leaves.

Life-cycle

Moths are stout, forewings are dark with wavy white markings. Hindwings are white with brown margins. Eggs are laid in clusters of 200-300 underneath the leaves and are covered with brown bristles. Incubation period is 4-5 days. Fully grown caterpillars are stout, cylindrical and pale greenish brown in colour, with dark markings. They have transverse and longitudinal grey and yellow bands. The caterpillar is nocturnal in habit, and lies hidden in crevices of loose soil at the base of the plant during day time. It measures 3.5-4 cm in length and pupates in soil. Larval period is 2-3 weeks. Pupae are dark brown in colour and measure 1 to 1.5 cm in length. Moth emerges from pupa in about 2 weeks. Life-cycle is completed in 36 to 40 days.

Management

Adopt Integrated Pest Management approach comprising of cultural/mechanical, chemical and biological control methods.

Cultural/Mechanical practices:

— Collect and destroy egg masses and young caterpillars.

— Plough the pest-infected garden and dig near the base of mulberry plant to expose the pest to sunlight and predators.

— Use light traps to attract moths by placing a tray with 0.5% soap solution below the light source.

Chemical Control:

— Spray 0.15% DDVP [2 ml DDVP (EC=76%) in 1 litre of water] in the evening, 20 days after pruning or leaf harvest. Feed silkworms 10 days after pesticide application.

— Dust 5% Malathion near the base of the plants immediately after pruning or leaf harvest.

Biological Control: Use 'Spodolure', a pheromone trap @ 2 lures/acre twice at an interval of 15 days, starting from 25 days after pruning or leaf harvest to attract and kill male moths.

"Tukra" (Maconellicoccus Hirsutus)

"Tukra" is caused due to infestation of mulberry plants by the mealy bug, Maconellicoccus hirsutus (green) (Hemiptera: Pseudococcidae). Though "Tukra" infestation is observed throughout the year, it is severe in summer.

Symptoms

Malformation of the apical shoot due to flattening and/or

thickening of the affected portion of shoot, reduction in internodal distance and wrinkling and curling up of apical leaves is observed. Affected leaves become dark green in colour and turn pale yellow, in the case of severe attack. Affected portion of plants become brittle. Presence of white mealy substance on infested portion of plants confirms mealy bug infestation. The mealy bugs suck the sap, resulting in stunted growth of the plant and reduction in leaf yield.

Life-cycle

Each gravid female deposits about 250 eggs. The eggs hatch in 5-8 days. The crawlers are orange in colour and this stage lasts for a day. They settle on the plant and secrete a white mealy substance on their body. Females have three nymphal instars, while males have four. Life-cycle is completed in 24-26 days. Adult males are slender and possess a pair of wings and caudal filaments, each. They predominantly reproduce parthenogenetically.

Management Practices

Adopt an Integrated Pest Management (IPM) approach comprising cultural/mechanical, chemical and biological methods.

Cultural/mechanical practices:

- Clip the affected portions of the plants, collect them in polythene containers and dip in 0.5% soap solution (5g of soap in 1 litre of water) to destroy the pest. This practice should be employed about a week before commencement of leaf harvest.
- Do not grow alternate host plants such as croton, hibiscus, guava, grapes, lady's finger, cotton, etc., in the vicinity of mulberry garden.

Chemical Control: Spray 0.15 DDVP [EC=76%) in 1 litre of water] in 0.5% soap solution on pruned or leaf harvested plants, twice at an interval of 10 days. Leaves can be fed to silkworms, 10 days after spraying the insecticide. This practice should be adopted during summer (February-May).

Biological Control: Release the predatory coccinellid ladybird beetle, Cryptolaemus montrouzieri (Muls) (Coleoptera: Coccinellidae) @ 125 adults per acre twice, during August-September and October-November.

Precautions

— Before pruning the infested mulberry plantation, clip the affected portions, collect in a polythene bag and dip in 0.5% soap solution.

— When cuttings from "Tukra" affected plantations are procured for planting, dip them in 0.15% DDVP solution for 2-3 minutes before planting.

— Ants play an important role in spreading "Tukra" infestation in mulberry plantations. Locate ant nests in and around mulberry plantations and treat them with 5% Malathion dust soon after pruning or leaf harvest.

Wingless Grasshopper, Neorthacris Acuticeps Nilgirensis

The wingless grasshopper, Neorthacris acuticeps nilgirensis Uvarov (Orthoptera: Acrididae) is a serious pest in rainfed mulberry plantations. The rate of infestation ranges from 0 to 95%. Besides causing extensive damage to mulberry, this pest infests sunflower, ragi, groundnut, beans, potato, etc. The pest occurs throughout the year and the infestation is severe during July-August.

Symptoms

Nymphs and adults feed voraciously on sprouting buds and leaves of mulberry. Sometimes, they also feed on green bark of affected plants. When the infestation is severe, mulberry plants are completely devoid of leaves.

Life-cycle

Adults of the grasshopper are light in colour. Gravid females lay 6-8 eggs pods in loose soil at a depth of 2-3 cm. Each egg pod contains 8-18 eggs. Eggs hatch in 28-31 days. The I, II and III instar nymphs are straw coloured, whereas the remaining instars are light green in colour. Nymphs pass through 6 instars in 90-95 days. Females are larger in size than males and live for 45-60 days. The pest completes its life-cycle in 163-186 days.

Management

Cultural/mechanical measures:

— Plough the infested mulberry plantations immediately after the onset of monsoon, to expose the egg masses to sunlight and predators.

— Maintain field sanitation by keeping the mulberry garden free from alternate host plants.

Chemical Control: Spray 0.076% DDVP [1 ml of DDVP (EC=76%) in 1 litre of water] on mulberry foliage to kill nymphs and adults. If infestation is severe, a second application is necessary, 10 days after the first spray. Harvest mulberry leaf for silkworm rearing 10 days after the last spray.

Insecticide Preparation

The active toxicant in a commercial formulation suitable for direct application is rarely available because small quantity of active ingredient has to be distributed on a large area. Dilution, therefore, becomes another important aspect for consideration. To obtain the weight or volume of solid or liquid formulation of commercial insecticide available in market for the preparation of a solid or liquid containing a desired (or recommended) percentage of toxicant, the following formula can be applied.

1) $d = (aXb)/c$

where

- a = % of toxicant desired (recommended).
- b = Wt (in g or kg) of dust or volume (in ml or lt) of liquid required for application.
- c = % of toxicant available in the commercial insecticide formulation.
- d = Wt (in g/kg) or volume (ml or lt) of commercial formulation required.

2) After obtaining the value of d, the following calculation formulation should be carried out to obtain the weight or volume of diluent (e),

$$\text{value of (b)} - \text{value of (d)} = e$$

3) Mix well the diluent (value of e) by weight (in case of solid formulation) or by volume (in case of liquid formulation) to commercial formulation (value of d) to obtain the required quantity of desired strength of particular insecticide.

Insecticides are widely used for the control of insect pests of several field crops. Improper use and their application results in undesirable drift residues to adjacent field crops. Accordingly, mulberry gardens have also been found contaminated due to use of insecticides in the nearby crops.

Since silkworms are highly sensitive to insecticides, harvest of mulberry leaves before the safe period should be completely avoided. In case, this is not followed scrupulously and contaminated leaves are harvested from the garden either located adjacent to field crops applied with insecticides or directly treated with insecticides, silkworm larvae develop toxic symptoms followed by the loss of the crop. These symptoms include:

— Vomiting of the digestive juice

— Swinging of the anterior half of the body

— Shortening of the body due to loss of the body fluid

— Muscle contraction and

— Paralysis is followed by death of the silkworm larva.

To avoid loss of silkworm crops (due to feeding of contaminated leaves), mulberry leaves should be harvested on or after the number of days shown against each insecticide given below.

Safe periods of various insecticides against silkworm

	Name of insecticide * (% active ingredient)	Concentration Safe period ** (in days)
Demeton	0.01	11
(Metasystox)	0.05	13
Demeton	0.01	07
(Metasystox)	0.05	11
Aldrin	0.01	11
(Aldrex)	0.05	13
Phosphomidon	0.01	11
(Dimecron)	0.05	13
DDVP	0.01	07
(Dichlorovos, Nuvan)	0.05	11
Methyl parathion	0.01	07
(Metacid)	0.05	13

Carbaryl	0.05	09
(Sevimol)	0.10	11
Quinalphos	0.05	07
(Ekalux)	0.10	13
Dimethoate	0.05	07
(Rogor)	0.10	11
Endosulfan	0.05	09
(Thiodan)	0.10	17
Phosalone	0.05	09
(Zolone)	0.10	11
BHC	0.05	07
(BHC)	0.10	11
Chlordane	0.05	11
(Termex)	0.10	13
Malathion	0.05	13
(Cythion)	1.00	17

* Names in parentheses are the trade names of insecticides
** Safe periods have also been indicated with each insecticide recommended for mulberry pest control.

Precautionary Measures

— Do not allow children for spraying

— Do not spray in hot sun, spray in cool hours

— Do not blow nozzle, use a needle or thin wire to clean

— Do not spray against wind, keep away from spray mist and dust drift

— Do not mix by hand, use rod for mixing

— Wash the hands or take bath before eating

Disinfectants and Disinfection Methods

Disinfection forms an integral part of healthy and successful silkworm rearing. It aims at the total destruction of disease causing pathogens. Several diseases caused by bacteria, viruses, fungi and protozoa affect the silkworms. These pathogens released by diseased silkworms easily accumulate and spread in the rearing environment through different routes. They are not easily destroyed and can persist/survive for long periods under congenial conditions. The spores of the pathogens, especially those of fungi are light and can easily be drifted by air current resulting in easy spread of diseases.

There are no curative methods for any of the silkworm diseases and they are best prevented than cured. This is achieved by adoption of proper and effective methods of disinfection and stepwise maintenance of hygiene during rearing. To realise the benefit of disinfection to the full extent, attempt should be made to have synchronised disinfection (mass) and rearing at village or block level considering them as one unit.

Chemical Disinfectants

Formalin

It is commercially available as 36% formaldehyde in solution form. A mixture of 2% formalin + 0.05% detergent is an effective solution that can be used for disinfection purpose as spray. Formalin is effective only in rearing houses, which can be made airtight and it is faster and more pronounced at temperature above 25°C and humidity more than 70%.

Bleaching Powder

It is a white amorphous powder, with a pungent smell of chlorine. For effective disinfection, a high grade bleaching powder with an active chlorine content of 30% must be used. It should be stored in sealed bags, away from moisture, failing which it will be rendered ineffective. The action of bleaching powder is optimal under wet contact conditions and therefore the surfaces of equipment and walls should be drenched with this solution. A 2% bleaching powder in 0.3% slaked lime solution is used for disinfection as spray.

Slaked Lime

A very useful bed disinfectant in sericulture, especially against viruses. It absorbs moisture and can be used to regulate bed humidity and maintain hygiene. Application of lime dust in combination with bleaching powder in an around rearing houses and premises improves hygiene in the environment.

Chlorine Dioxide

Chlorine dioxide marketed as Sanitech is an ideal disinfectant for sericulture. The disinfectant available at 20,000 ppm concentration is a strong oxidizing agent, effective at broader pH range and at 2.5% concentration in combination with 0.5% slaked lime is effective against all silkworm pathogens. It is stable and may be activated at the time of its use. It possesses tolerable odour and least corrosive at the suggested concentration.

Required Materials for Disinfection

Disinfectants, detergent, sprayer – Rocking or power sprayer, buckets, measuring jar, weighing scales, gas masks, metal pans, room heaters, slaked lime powder. Hand gloves and muslin cloth.

Prevention Technologies

The technology for prevention/control of silkworm diseases during silkworm rearing is provided as sequential steps.

Cleaning of Rearing House and Appliances

Immediately after the disposal of cocoons, collect at one place inside the rearing house, all diseased and dead larvae, pupae, floss, left over mulberry, bed refuse, silkworm faeces, dust, dirty, etc., and disinfect by sprinkling 5% bleaching powder solution and dispose off by burying at 2ft. depth or burning.

Disinfection of Rearing House and Appliances

Measure the length and breadth of the rearing including leaf preservation room, mounting room/place, etc. and calculate the floor area for disinfection.

Floor area = Length x Breadth (of the floor)

Example: Length of floor 20′, breadth of floor 15′

Area of floor = 20 x 15 = 300 sq. ft. = 28 m^2

The disinfectant required for disinfection of rearing house is 2 lt/sq.m floor area or 185 ml/sq. ft. floor area. To estimate the quantity of disinfectant solution required, multiply, the floor area in square metre by 2.0 or the floor area in sq. ft. by 0.185 which gives the actual disinfectant required in litres for disinfection of inside or rearing house.

Disinfection

Disinfect the rearing house using 2% bleaching powder in 0.3% slaked lime, 2.5% Sanitech (Chlorine dioxide) in 0.5% slaked lime or 2% formalin + 0.05% detergent solution.

Prepare the disinfectant solution using highgrade stable bleaching powder containing 30% chlorine and having ISI mark. Always prepare disinfectant solution prior to disinfection. To prepare the solution, determine the quantity of disinfectant solution required as suggested above and calculate the quantity of ingredients to be added. To prepare 1 lt. Of 2% bleaching powder in 0.3% slaked lime solution, the requirement is 1 lt. Of water, 20 g bleaching powder and 3 g slaked lime powder. Make a paste of 20 g bleaching powder and 3 g of slaked lime in 100 ml of water and to this paste, add water to make 1 litre solution.

Spray using powerful jet sprayer, the required quantity of disinfectant @ 2.0 lt/m floor area of rearing house + 10% for outside of rearing house +25% of disinfectant solution for appliances) uniformly to drench all parts of rearing house inside and outside, and appliances. Keep the rearing house closed for a minimum period of 6-10 hours. Disinfect bamboo mountages with 2% formalin solution. Do not spray bleaching powder solution to the mountages. After minimum period of 18-24 hours, shift the rearing appliances out of rearing house and sundry for 10-12 hours. Do not smear cowdung to the rearing trays and store them inside the rearing house till 5 days prior to next brushing.

Disinfection Prior to Brushing

Disinfection process should start 4-5 days prior to brushing. However, the eggs may be incubated in a separate disinfected incubation room.

Five days before brushing

The rearing house and appliances are cleaned, washed in water. If trays and other appliances namely, basins, vinyl sheet, leaf basket, etc., are used, conduct additional disinfection by dipping them in disinfectant or spraying disinfectant.

a) *Dipping in disinfectant*: Disinfect the rearing appliances that could be disinfected with 2%

bleaching powder in 0.3% slaked lime solution by dipping them for 10 minutes in the solution in a disinfection tank. A tank of 2 feet depth and 4 feet diameter is suitable for disinfection. Prepare the disinfectant solution to fill half of the height of the tank.

b) *Disinfection by spraying*: Disinfection of appliances by dipping in disinfectant is the best method. However, where the facility of disinfection tank is not availabie, the following method is advised. Disinfect the appliances using 2% formalin + 0.05% detergent solution or 2.5% of Sanitech in 0.5% slaked lime or with 2% bleaching powder in 0.3% slaked lime solution. Spray the disinfectant @ 35 ml/sq. ft. surface area or 700 ml for a tray of 3.6 feet diameter. After the spray, put them together and keep completely covered with vinyl sheet for a minimum period of 6 hours. Disinfect the mountages also, following the above method.

Four days before brushing

Sun dry the appliances. If the prevalence of viral diseases (Grasserie and Flacherie) were high during the previous crop, spray 0.3% slaked lime in water (3g/lt) to the rearing house and appliances @ 2 lt/sq.m floor area + additional requirements. Sun dry the appliances after 1-2 hour of spray.

Three days before brushing

Conduct second disinfection of rearing house and appliances. Shift all the disinfected appliances into the disinfected rearing house and arrange in the room. Disinfect the rearing house and appliances by spraying 2% bleaching powder in 0.3% slaked lime solution or

2.5% Sanitech in 0.05% slaked lime or 2% formalin + 0.05% detergent solution. The quantity of disinfectant solution required is calculated and sprayed as per the first disinfection. After disinfectant spray, keep the room closed preferably for 24 hours. The second disinfection may follow the first if the gap between the two is not much.

Two days before brushing

Dust 5% bleaching powder in slaked lime powder @ 200 g/sq.m at the entrance of the rearing house and the passage to it. Sprinkle water @ 1 lt/sq.m floor area. Open the windows of the rearing house and ventilate to drive off all the adour of disinfectant. Prevent the entry of unauthorised persons into the rearing room to avoid contamination of rearing house and appliances.

Estimation of requirement of disinfectant dust for floor disinfection:

Floor area (sq.m) x 0.200 = kg of dust required for dusting

Floor area (sq.ft) x 0.018 = kg of dust required for dusting

Preparation: 5% bleaching powder in slaked lime powder.

Preparation of 1 kg: Mix 50 g of bleaching powder in 950 g of fine slaked lime powder.

One day before brushing

Arrange appliances for chawki rearing and maintenance of hygiene. Keep the later instar rearing room and mounting hall closed and open only one day before the later instar larvae are to be shifted into the rearing house.

Disinfection and Maintenance of Hygiene

— Avoid borrowing the rearing appliances. Do not use appliances without disinfection. Avoid overlapping rearings.

— Maintain personal and rearing hygiene throughout the rearing. Special consideration for maintenance of hygiene be given during the chawki stage. Restrict the entry of persons into the rearing house.

— Surface disinfects the silkworm eggs before head pigmentation stage with 2% formalin for 5-10 minutes and wash with water. Dry them in shade and incubate following the guidelines.

— At the passage to entrance of rearing house, sprinkle 5% bleaching powder in slaked lime @ 18 g/sq.ft. An area of 6 x 8 feet may be covered.

— Rearer and any person entering into the rearing house must disinfect foot and hard before entry. These persons should step on foot mat at the entrance containing a mat soaked in 2% bleaching powder in 0.3% slaked lime solution. Ultimately a disinfection tank of 2 x 1½ ft. x 3 inch may be made and filled with 2% bleaching powder in 0.3% slaked lime solution. Replace the disinfectant everyday.

— Wash hands with two 2% bleaching powder in 0.3% slaked lime solution or "Suchi" and then rinse with clean water. Replace the disinfectant everyday.

— Pick up diseased/unequaled/suspected disease worms and disposed into 5% bleaching powder in slaked lime in basin. Keep in basin covered with a lid and change every 3-4 days.

— Disinfect the hand after picking the diseased worm by washing in 2% bleaching powder in 0.3% slake lime solution or by "Suchi" solution.

— Clean the silkworms rearing bed using bed cleaning net and disinfect after every use by dipping in 2% formalin solution for 10 minutes. Wash hands with disinfectant.

— Spread vinyl sheet for collection of bed refuse and shift bed refuse into manure pit. Disinfect the vinyl sheet after use every day by dipping in 2% bleaching powder in 0.03% slaked lime solution.

— Rare silkworms always on rearing seat paper (newspaper). Change as and when soiled.

— Wipe the floor after cleaning with 2% bleaching powder in 0.3% slake lime solution or by "Suchi" solution. If the floor is a mud floor, dust 5% bleaching powder in slaked lime at an interval of 3-4 days.

— Dust the bed disinfectant as per the schedule, quantity and precaution. Feed 30 minutes later. While extending the bed also, apply bed disinfectant to the empty tray.

— Do not store mulberry leaves inside the rearing room. Store in a separate disinfected room.

— Feed good quality and sufficient quantity mulberry leaves for robust larval growth. Provide recommended spacing sufficient ventilation for healthy growth of later instar worms.

— Avoid contamination of mulberry. Use separate baskets for collection of mulberry and feeding worms.

— Avoid injury to silkworms while rearing and mounting for spinning cocoons. Sprinkle bleaching powder in mounting place.

— Pick diseased and non-spinning larvae from mountages and dispose them into 2% bleaching powder in 0.3% slaked lime solution in a basin. Wash hands and feet while coming out or rearing house/mounting hall, etc. After completion of work at every visit.

Disinfection of Silkworm Body and Rearing Seat

It is most important to disinfect the silkworm body and rearing seat with bed disinfectants for containing secondary contamination and to improve the hygiene during rearings.

Precautions

— Do not dust on the silkworms settled for moult or under moult.

— Do not dust on the silkworms where eatable mulberry leaves are available.

— Dust uniformly as per schedule, quantity and with precautions. Feed after 30 minutes.

— While dusting, protect eyes, nose, ear and mouth. Wash hands thoroughly after dusting bed disinfectants.

— Keep disinfectants away from children.

Silkworms Diseases

Muscardine and Apergillosis

Muscardine and Aspergillosis are fungal diseases of silkworm. The diseases prevail during winter and rainy season in all sericultural areas. The chawki rearing conditions are congenial for aspegillosis. White muscardine is caused by a fungus, Beauveria bassiana. Apergillus is caused Aspergillus flavus, A. oryze and A. tameri.

Symptoms

The larvae loose appetite, become inactive and on death, gradually, they become harder and finally mummifies. The fungus overgrow the larvae, develops fruiting body and produce conidia on the surface giving white colour to the mummified diseased silkworm. (White muscardine). Aspergillosis generally infect early instar silkworm. The early instar silkworm becomes inactive and dies without clear morphological symptoms. Late instar silkworm develops black spot at the site of infection and dies due to Aflotoxin by the fungus.

Predisposing factors

Low temperature and high humidity to white muscardine. High temperature and high humidity for Aspergillosis.

Source of pathogen

Mummified/diseased silkworm, contaminated rearing environment, rearing house, appliances and alternate hosts (Most lepidopteran pests). Aspergillus sp. are common in the environment growing as a saprophyte.

Transmission

The conidia of the pathogen are dispersed by wind or contact. The conidia on contact with host integument, germinate, penetrate into the body cavity and cause disease. The contaminated rearing environment, house, appliances and alternate hosts also form source for the spread of the disease.

Prevention

— Practice disinfection of silkworm rearing house, surroundings, appliances and silkworm egg surface. Aspergillus sp. are comparatively more tolerant to formalin hence, 3% formalin solution is suggested.

— Practice hygienic measures during silkworm rearing. Ensure the measures for destruction of diseased silkworm.

— Regulate humidity by dusting slaked lime at the time of every moult and in between, if necessary to keep the rearing bed dry. Provide good ventilation.

— Apply bed disinfectant as per recommended scheduled and quantity.

— As specific measure, dust 1-2% for 3rd, 4th and final instars) of Dithane M45 in Kaolin or Captan in slaked lime on silkworm body immediately after every moult and on the 4th day final instar @ 3-5 g/sq. ft. Alternately, apply formalin chaff to surface disinfect silkworm body. Formalin chaff is prepared by partially burning paddy husk and mix formalin of concentration of 0.6%/0.8% (0.6% formalin chaff is for chawki worms and 0.8% for later stages) in the ratio of 10:1. The chaff is sprinkled uniformly on the worms after every moult and paraffin/old newspaper is covered for 30 min. Feeding should be given afterwards.

— Feed good quality of mulberry, provide requisite spacing and ventilation. Maintain the required temperature and humidity.

Nuclear Polyhedrosis (Grasserie)

Nuclear polyhedrosis is a major viral disease commonly known as Grasserie or Haluthode. The disease prevail although the year but is insentive during summer and rainy seasons.

Causative agent

Bombyx mori nuclear polyhedrosis virus, a baculovirus.

Symptoms

At the Early Stage of Infection: The infected silkworm appears normal and microscopic examination of the haemolymphy may indicate the presence of nuclear polyhedra with in the nucleus of haemocyte.

Late Stage of Infection: The skin becomes shining before moulting and fails to moult. The integument will be fragile and the intersegmental region becomes swollen. The haemolymph becomes tubid white and contains large number of polyhedra which are hexagonal.

Infection

Silkworm gets infected when it feeds on contaminated mulberry leaves.

Sources of infection

Disintegrating diseased silkworm, its body fluid, alternate hosts and contaminated silkworm rearing house and appliances.

Predisposing factor

High temperature, humidity and poor quality of mulberry.

Transmission

The disease is transmitted by secondary infection of larvae feeding on the contaminated mulberry.

Spread of disease

The diseased silkworm extrudes the pathogen along with

oozing body fluid during to injury and breakage of dead diseased larvae. The body fluid and dead broken larvae contaminate the rearing house, appliances, surrounding and mulberry. The disease spreads to the healthy silkworm on feeding the contaminated mulberry.

Prevention methods

- Practice disinfection of silkworm rearing house, surroundings, appliances and silkworm egg surface. A spray of 0.3 % slaked lime solution in addition to usual disinfection procedure is recommended for rearing house and appliances in the case of high incidence of disease.
- Practice hygienic measures during silkworm rearing. Ensured the measures for destruction of diseased silkworm.
- Identify and picked out infected larvae in the early stages.
- Apply any bed disinfectant as per recommended schedule and quantity.
- Maintain optimum rearing temperature and avoid drastic changes in temperature.
- Feed good quality of mulberry in sufficient quantity.

Flacherie

Flacherie is a syndrome associated with Infectious Flacherie, Densonucleosis, Cytoplasmic polyhedrosis, Bacterial diseases and Thatte roga in silkworm. The disease prevail although the year but is insentive during summer and rainy seasons.

The disease is caused by Bombyx mori virus (infectious Flacherie) and Bombyx mori densonucleosis virus (Densonucleosis) and Kenchu virus. Flacherie may be caused by these viruses individually as well as in association or each in association with Streptococci/ Staphylococci bacteria (Thatte roga). Cytoplasmic polyhedrosis is caused by a Reovirus, Bombyx mori cytoplasmic polyhedrosis virus. Streptococci sp., Streptococci sp. Basillus sp. And Serretia bacteria, individually and in association cause bacterial flacherie.

Symptoms

At the Early Stage of Infection: Symptoms are typical at the early stage of infection. Larva becomes lethargic and loose appetite.

At the Later Stage of Infection: The Larva becomes soft and flaccid. Retards in growth, becomes dull, vomits gut juice and faeces becomes sort with high moisture content. The cephalothoracic region becomes translucent (Infectious Flacherie, Densonucleosis and Cytoplasmic polyhedrosis). The midgut becomes turbid white due to formation of polyhedra and the faeces whitish black (Cytoplasmic polyhedrosis). The larvae ferments, turn different colour and smell foul (Bacterial diseases). Sometime symptoms of toxicosis are observed (Bacterial toxicosis).

Silkworm gets infected when it feeds on contaminated mulberry leaves. Sources of infection are: dead/diseased silkworm, its faecal matter, gut juice, body fluid, alternate hosts contaminated rearing house and appliances.

The disease is transmitted by secondary infection of larvae feeding on the contaminated mulberry. The diseased silkworm extrudes the pathogen along with gut juice and faecal matter throughout the incubation period of infection. The disintegrating diseased larvae also extrude pathogen into the environment. The pathogen contaminates the rearing house, appliances, surrounding and mulberry. The disease spreads to the healthy silkworm on feeding the contaminated mulberry.

Prevention methods

— Practice disinfection of silkworm rearing house, surroundings, appliances and silkworm egg surface. A spray of 0.3 % slaked lime solution in addition to usual disinfection procedure is recommended for rearing house and appliances in the case of high incidence of disease.

— Practice hygienic measures during silkworm rearing. Ensured the measures for destruction of diseased silkworm.

— Identify and picked out infected larvae in the early stages.

— Apply any bed disinfectant as per recommended schedule and quantity.

— Pick out diseased as early as possible and destroy them in disinfectant or by burning.

— Feed good quality of mulberry, provide requisite spacing and ventilation. Do not feed wet mulberry. Provide good environment. Maintain the required temperature and humidity.

Silkworm Pests

USI FLY, Exorista bombycis

The uzi fly, Exorista bombycis (Louis) (Diptera: Tachinidae) is a serious endo-larval parasitoid of the silkworm, Bombyx moriL. The pest was introduced accidentally into Karnataka in 1980. Since then, the pest has been inflicting considerable loss to sericulture industry. At present, he loss is estimated to be ranging from 10 to 20%. The pest occurs throughout the year. Its incidence is maximum from August to November.

Symptoms

Presence of black scar on the body of parasitised silkworms and maggot emergency hole in cocoon indicate uzi infestation. The silkworms parasitized in early stars are killed before attaining spinning stage, while those parasitised in the late V instar spin cocoons of weak built and from such cocoons uzi maggots emerge by piercing, thus rendering cocoons unfits for commercial reeling.

Life-cycle

Every gravid fly lay 1-2 eggs on each silkworm. The eggs hatch in 2-3 days and young maggots pierce into the host body and devour the body tissues. The maggots after spending 5-8 days of parasitic life, emerge from the host body. After emergence they lead 12-20 hours of post-feeding life and pupate in the loose soil, crevices on the rearing house floor, dark areas on the floor in the rearing house. The pupal stage last for 10-12 days the life cycle is completed in 18-24 days.

Prevention methods

Cultural/mechanical:

- Collect and destroy the uzi fly infected silkworms, uzi maggots and pupae.
- Keep the rearing house free from cracks and crevices.
- Uzi infected larvae spins cocoon a day or two earlier than the other silkworms. Such cocoon should be harvested separately and stifled.

Exclusion Method: Fix wire mesh or nylon net to the windows of the rearing house with an arrangement for ante-room at the entrance of the rearing house.

Chemicals

Uzitrap: It is a chemotrap used for attracting and killing the adults. Dissolve one tablet in one litre of water and the solution should be taken in light coloured flat trays or plates (10" diameter, or 10" Lx 8" B) ad place near the window (at the height of their base) both inside and outside the rearing house. The window needs to be provided with nylon net/wire mesh. In case there is a nylon net enclosure in the rearing house, on or two containers with uzitrap solution should be placed. The solution should be used starting from III instar of silkworm through spinning stage. If dust, dead insects, etc., accumulate in the trap solution, fresh solution should be prepared and used. One tablet of uzitrap containing 12 tablets is sufficient for 100 dfls rearing.

Uzicide: It is a liquid ovicidal formulation. It is sprayed on the body of the silkworms starting from the second day of the the 2nd day in III instar through 4th or 6th day in V instar on alternate days except the period of moulting. Uzicide should be sprayed after bed cleaning or 2-3 hours after feeding. Silkworms should be fed half an hour after spraying uzicide. Four to five litres of uzicide is required for rearing silkworms from 100 layings.

Uzi powder: It is an ovicidal dust formulation. It is dusted on the body of the silkworm on 2nd during III instar, 2nd and 4th days during IV instar and 2nd, 4th and 6th days in V instar. Uzi powder should be dusted after bed cleaning and silkworms should be fed half an hour after dusting. Four to five litres of uzi powder is required for rearing silkworms from 100 layings.

8

Technics of Silk Production

Today in Greece and the Balkans, labour intensive processes have been radically transformed following the introduction of mechanisation and the interbreeding of moths. Before World War II in Greece, raw silk was characterised by a high gum content: small goldenyellow cocoons produced low yields of degummed, golden-yellow raw silk. The equivalent cocoons today are large and white, and they contain little waste gum. This results in a far higher yield of raw silk. Within little more than half a century, raw silk production has been transformed beyond recognition. The same has happened in the Balkans. For this reason, it is generally unsatisfactory to compare contemporary Greek and Balkan practices with those of Byzantine times. Only the nonmechanised silk industries of rural China and India remain unaffected: they act as time capsules that provide living evidence for the intricacies of ancient silk production.

An unfortunate absence of relevant Byzantine sources concerning the essential processes renders it necessary to turn to non-Byzantine documentation. Chinese sources are particularly relevant, because sericulture was transposed to the Mediterranean from China. A key surviving Chinese agricultural treatise, the Keng tschi tuof A.D. 1149 provides a detailed account of moricultural, sericultural, and silk yarn-producing techniques.

Planting, Cultivation, Leaf Harvesting, and Pruning

The Keng tschi tu indicates that four main activities were essential for successful mulberry growing. In China, mulberries were planted on both flat and terraced land. The first question that arises is how were they grown in Byzantium? There seems to be no documentary evidence regarding this issue. However, if widespread cultivation across mountainous regions of Asia Minor is to be envisaged, the question of terraced mulberry plantations becomes important.

The same Chinese treatise indicates that mulberry seeds planted in spring took about a year to grow into saplings ready for transplanting. The saplings grew to maturity in fifteen years. Byzantine mulberry cultivation in mid-eleventh-century Calabria, as described in the Reggio Brebion, the land register of the Byzantine metropolis of Reggio, discussed below, indicates that only mature trees were taxable.

The Keng tschi tu emphasized an interrelationship between the nature and size of the mulberry grown and the time and frequency of subsequent leaf harvest. Evidently a continuous supply of fresh mulberry leaves could be ensured only with very careful planning. Timing and rotation of mulberry leaf harvests dictated the number and frequency of silkworm crops reared. It was not simply a matter of a single crop a year.

In Byzantium one knows that in the tenth century there was reliance on foreign imported raw silk, and this, together with the strict regulations against exporting Byzantine raw silk, suggests an overall scarcity of raw silk supplies. In light of the Chinese evidence about multiple crops of silkworms, one wonders why Byzantium could not independently supply its own industry. Was the production of raw silk too restricted and too highly taxed to attract sufficient investment? Alternatively, was local production simply too disrupted because, politically speaking, the plantations were sited in particularly vulnerable regions (first in Syria, later in Asia Minor)?

The Keng tschi tu indicates that the mulberry required well-dug soil, preferably irrigated. One may inquire how far the development of irrigation systems might have affected Byzantine mulberry cultivation. Well-irrigated plantations would have been larger, as irrigated trees needed to be more widely spaced. Such plantations would also have significantly increased the volume of raw silk production. Greater production in turn would have meant a larger industry. The growth of sericultural activity in the provinces in the eleventh to twelfth centuries, discussed below, could perhaps best be considered in relation to the development of irrigation systems in Byzantium in general. A. Harvey has indicated that in Byzantium, between the ninth and the eleventh century in particular, considerable interest was shown in the construction of elaborate-irrigation systems.

The Keng tschi tu clearly shows that in China sericulture was an ancillary occupation of farming communities. The farmers were the silkworm breeders. To what extent was this also true in Byzantium, and to what extent did landed magnates invest in mulberry plantations? One does know that, in the tenth century, wealthy individuals in Constantinople were permitted (within their own homes) to manufacture silks for personal use.

The Keng tschi tu indicates that the demands of raising silkworms were even more exacting than those of tending mulberry plantations. Large silk worm crops in mediaeval China were the responsibility of commercial breeders. A Chinese source of the twelfth to thirteenth centuries entitled the Farmers' Essentials detailed the precise activities of commercial breeders working in a thirteen bay silk worm rearing house.

At what point did commercial breeders appear in Byzantium? In the sixth century the private silk industry was dependent on state-controlled, imported raw silk supplies. By the tenth century, raw silk was produced in the hinterland, and imported raw silk was commercially available in Constantinople. The rise of Byzantine commercial breeding enterprises in the tenth century can be seen as heralding an increasing decentralisation of raw silk supplies in the eleventh to twelfth centuries.

The Byzantine commercial breeders must have adhered to rules like those set out in the Keng tschi tu and in other Chinese treatises. These reveal that, after shedding their skins either three or four times according to their breed, silkworms were ready to spin twenty-eight to thirty-five days after hatching. The Chinese silkworms described in the Farmers' Essentials were set on suspended trellises and encouraged to spin by warming. The Chinese Book of *Sericulture of Sun Kuang-Hsien* (died 968) revealed how warming also prevented the silkworms from wandering while they were spinning their cocoons.

It is clear from the Chinese documentation that, after spinning, cocoons were either stored in salted jars, stifled by cooling, or speedily unraveled according to need. The question is in what form did Byzantium receive its imported raw silk? The Book of the Eparch indicates that both domestic and foreign imported raw silk was reaching Constantinople in the tenth century. Although the transport of salted cocoons from afar would have been a possibility, it would have been more satisfactory for the silk to have arrived in the form of unraveled and reeled silk. Salted silk cocoons would have been brittle and would have yielded poorer silk yarn.

Foreign cocoons would also have been bulky to transport over long distances and unprofitable to carry (especially in relation to their raw silk yield and selling price). On the other hand, locally produced cocoons would not have had high transportation costs as they would not have had far to travel, and they would not have required salting. They could have been swiftly unraveled by appropriate members of the silk guilds of Constantinople, as discussed below. It is most plausible to suggest that both cocoons and unraveled raw silk were arriving in Constantinople in the tenth century. But it is more likely that local unsalted cocoons would have been traded rather than foreign salted cocoons. The exclusion of brittle, foreign, salted cocoon yarn would have protected the high reputation of the silk industry of Constantinople.

Silk Yarn Manufacturing

The production of the silk yarn was a laborious and specialised activity. In medieval China, as illustrated in a handscroll datable ca. A.D. 1200-1210, special reeling stoves were evolved (compare those of rural India today). These stoves consisted of a basin above a fire. The cocoons were floated in the basin in a bath of heated water, and their surface was teased to loosen the end of the silk thread of each cocoon. The individual silk starting threads were then fed through separate hooks and led onto a reeling device. Different numbers of cocoons were reeled simultaneously to produce silk yarns of varying weights. The medieval Chinese silk yarns ranged from very fine (the equivalent of see-through silk stockings today) to heavy silk furnishing fabric quality. Surviving Byzantine silks also use very variable weights of silk yarn, which would argue for the early existence of reeling basins in Byzantium.

During reeling, little twist could be added to the threads, and to strengthen silk yarn intended for warps, extra twist (up to 2,000-3,000 turns per meter) had to be added. In China, first the spindle whorl and then, by the tenth to the eleventh century, the spindle wheel were developed for the purpose of adding twist to silk threads. In Byzantium, judging by the high twist on warps of surviving silks, there can be no doubt that some form of spindle wheel existed by the tenth to eleventh century.

Prokopios argued that imperial regulation of silk prices had a ruinous effect on the provincial nonimperial silk industry. R. Lopez interpreted Prokopios' account to mean that the emperor had set a ceiling price of 8 nomismata per pound for the purchase of imported raw silk, but N. Oikonomides took this to refer to silk garments. This low price meant that foreign producers would not sell. If this was true, how did the imperial silk industry survive? Were there imperial mulberry plantations and imperial sericultural establishments to supply the imperial industry? If so, where were these situated? Is it possible that the imperial estates in Syria under the care of Magnos the Syrian, the kommerkiarios (silk official), included mulberry plantations? Without further documentary evidence, this is uncertain, but it must remain a possibility.

The Peri Metaxes sets a ceiling price of 15 nomismata per pound for raw silk. The problem is to reconcile the 8 nomismata ceiling set for raw silk/woven silk garments described above with the ceiling price stipulated by the Peri Metaxes. The difficulty lies in dating the Peri Metaxes. Some scholars have suggested that it can be dated before 540, under Justinian, but Oikonomides suggested that the 15 nomismata ceiling price may have acted as only a nominal figure and that it is difficult to date.

These ideas need to be balanced from a practical point of view. If raw silk cost a maximum of 15 nomismata per pound at the time of the Peri Metaxes, one can envisage that a lightweight shift dress (2 pounds in weight) could be woven for around 30 nomismata. On the other hand, if 8 nomismata was the ceiling for silk garments as against raw silk, the garments in question at a time contemporary to the Peri Metaxes would have to have been very lightweight indeed. Only a very flimsy dress could be produced from half a pound of silk. Surviving sixth-to-seventh-century Byzantine silks are all relatively heavyweight textiles: all would have weighed far more than half a pound if made into garments.

Concerning the distribution of mulberry plantations and the production of raw silk, the evidence of the seals of the kommerkiarioi is important. The earliest seal of a kommerkiarios belonged to an officer based in Antioch under Emperor Anastasios (491-518). It is evident that the kommerkiarioi originally acted for the state but that subsequently they could draw their own profits. The seals reflect the main thrust of sericultural activity across the

Byzantine Empire up to the twelfth century: sericulture evidently passed from Syria to Asia Minor and then into the Balkans.

There were five stages of raw silk acquisition up to the twelfth century: (1) an initial phase centered in Syria before the fall to the Arabs (5th-7th centuries); (2) a subsequent stage of sericultural activity within Asia Minor (8th-9th centuries); (3) a third period of activity concentrated in western Asia Minor and the Balkans (9th-10th centuries); (4) a further initiative that saw the importation of Syrian silks to boost domestic supplies (10th century); (5) finally, a decentralisation of raw silk supply (11th-12th centuries). Provincial Byzantine raw silk (Sicilian and Calabrian) appeared on the market at Fustat. Italo-Byzantine, provincial Byzantine silk from Asia Minor as well as imported Islamic raw silks may have been available for use by weavers in the Peloponnese.

The general picture that emerges between the fifth and the twelfth century is one of selective expansion and of increasing decentralisation. It is insufficient just to chart these developments. From a practical point of view, these changes in raw silk supply had many ramifications for Byzantine weavers. The quality and nature of silk yarn would have governed both weaving technique and ease of production of designs current at any one time. The slightest change in yarn supply affected both the preparation of the loom and the subsequent execution of the design. Such changes of raw silk supply had to be accommodated by skilled weavers who could envisage how techniques and designs had to be adapted to fresh supplies of yarn.

The fact that five major shifts of yarn production can be discerned, and that domestic supplies also had to be supplemented by imported supplies, indicates that Byzantine weavers must always have been a highly skilled workforce. The specialist divisions of labour as described in the Book of the Eparch encouraged the maintenance of high standards.

It is difficult to pinpoint the exact location of mulberry plantations before the year 1200. The first Byzantine source that specifically details a substantial mulberry plantation is the Reggio Brebion. In the Reggio Brebion, mature mulberry trees (i.e., those more than fifteen years old) were subject to imperial taxation at the rate of 2,436 taria (4 taria to the dinar). It is not stipulated whether or not the mulberry leaves were fed to silkworms, although the existence of extensive moricultural and sericultural activity in Calabria in later times would suggest that silk production was involved.

Looms and Weaving Techniques

Weaving techniques impose limitations on the types of designs that can be woven on silks. In turn, weaving techniques are dependent on the types of looms used. In Byzantium it was necessary to devise looms with special pattern-producing devices (pattern harnesses) to accommodate increasingly intricate pattern motifs. But sophisticated looms were of no use without skillful weavers. Manual dexterity had to keep pace with technological development to produce silks of the high quality demonstrated by the surviving Byzantine textiles.

The main weaving techniques found on the surviving Byzantine silks dating before the thirteenth century are: tabby, damask, twill, lampas, and tapestry weaves. (Satin weave, in which the horizontal thread or weft was permitted to pass over four or more vertical threads or warps before being bound down, was not developed until the 13th to 14th century). It is useful to define these five weaving types in conjunction with the evidence of surviving examples.

Tabby Weave

In tabby weave the horizontal or weft thread is passed alternatively over one and under one vertical or warp thread. All the warps that lie above the weft in the first pass of the thread across the loom lie below it in the next, above it in the third pass, and so on. Tabbies may have either one or two systems of warps and wefts. In tabbies with a second warp, this is hidden between the upper and the lower surface of the weave, and it acts merely to guide the weft to the obverse or the reverse of the fabric. Some early tabbies with a single warp and weft also employed floats of weft threads for patterning effects. Here the wefts are floated over a tabby to produce simple geometrical designs.

Damask Weave

Damask is a weave with a single warp, and the fabric is reversible. The threads are bound in twill. There are two faces to twill binding. Where the weft predominates, a weft-faced weave results; but where the warp predominates, a warp-faced twill is formed. The damask contrasts the warp and the weft faces of twill binding.

Twill binding itself, in the case of weft-faced twill, means that the weft is passed over two or three warps and then under one warp, over two or three warps and under one, and so on from one side of the loom to the other. Each successive row begins one warp further in, creating a diagonal furrow down the silk as weaving progresses. The same binding occurs using the warps instead of the wefts in warp-faced twill binding. Damasks are monochrome weaves that rely on changes in weave rather than on color contrasts for the formation of the pattern.

Twill Weave

Twill binding, as described above, is used for this weave. There are two warps: a binding warp that secures the weft at required intervals, and a main warp, which sits between the two faces of the weave but does not appear either on the surface or the obverse of the weave. The main warps are lifted or lowered according to the needs of the design, and the action of the main warps determines the correct opening of the sheds for the weaving of the patterns. Main warps are either single or paired in Byzantine silk twills. They are of degummed silk twisted to the right (Z).

Central Asian silk twills of the seventh to tenth century used main warps grouped in three to fours and twisted to the right (Z). Later Central Asian twills used paired, gummed silk warps that did not require twisting. They often imitate Byzantine designs but are easily distinguishable from Byzantine twills. Byzantine silk manufacture appears to have been largely dominated by twill weaves. Twills were most often polychrome, but in the tenth to eleventh century a fashion also emerged for monochrome twills on which the designs appeared through a change in weave rather than by virtue of color contrasts.

Lampas Weaves

Lampas weaves were developed in both Byzantium and the Islamic Mediterranean around the year 1000. In the Islamic world they were widely woven in Spain and Iran. They are monochrome silks that rely on changes in weave rather than on color contrasts for the formation of their patterns. In early Byzantine lampas weaves, the main warps did not help to bind the wefts, but in developed lampases they did do so.

Two types of developed lampases can be distinguished: tabby, tabby lampas weave and tabby, twill lampas weave. In tabby, tabby lampas weave, the two faces of tabby binding (weft and warp faces) are contrasted to create the design. In tabby, twill lampas weave, tabby and twill bindings are contrasted to delineate patterns.

The main warps in lampas weaves are grouped in sets containing combinations of single, paired, or tripled main warps. The groupings of such warps may serve to characterise certain groups of silks (e.g., Spanish lampases as distinguished by D. Shepherd). Tapestry Weave Tapestry weave silks do not survive in large number, but Byzantine examples show the use of the slit tapestry technique. Here each color area is separated fromthe next by a slit in the weave. The slits occur where the wefts of one color are turned back upon themselves rather than being carried over into the next color area of the design.

Medieval Looms

A history of the Byzantine hand draw-loom has not yet been written. Surviving pictorial representations of Byzantine looms do not provide any idea of the complexity of the actual looms that must have existed to produce the surviving Byzantine silks. Simple horizontal looms are depicted with shafts but without any form of developed pattern-producing device or "figure harness" that would have been essential for weaving the intricate Byzantine patterns that survive.The Byzantine looms must be envisaged with reference to the surviving silks and also in conjunction with documentary evidence for advanced hand draw-loom weaving in Chinese sources.

J. Becker has suggested that a number of stages can be discerned in what he termed the development of "mechanical patterning" (i.e., in hand draw-loom weaving). He suggested a move from the use of pattern-weaving devices or pattern heddle rods (sticks inserted into the warps and lifted as required) to true pattern shafts (devices used to create the correct sheds through which the weft was passed). Next he envisaged the development of a form of cross harness whereby strong cords were attached to individual warps across the loom by way of heddle loops. The cords were then knotted to a vertical draw string and suspended above the loom. The draw strings were pulled in the correct order for the opening of the sheds required for the creation of more complex designs. This form of loom is still operational in India today, but it does not seem to have been characteristic of medieval China.

In China there appears to have been a hand draw-loom with a pattern harness placed in a frame behind shafts on the loom. Looms of this type, with two shafts and with a draw harness and individually weighted harness cords, are depicted on a Chinese scroll of the Sung period (960-1277). A reconstruction of this type of loom was exhibited in Belgium in 1989, and its construction was based on a nineteenth-century loom from Beijing. The 1989 loom did not have a device for the automatic repeat of the pattern, as did the Beijing loom. Byzantine hand draw-looms after the tenth century probably did include a device for the repeat of patterns, judging by the evenly sized repeats of important extant silks.

A developed hand draw-loom with pattern-repeating device must be envisaged for Byzantine silks such as the Aachen Elephant fabric of the early eleventh century. This used asymmetrical motifs more than a meter wide enclosed in foliate medallions. Each design required 1,440 manipulations of a pattern-producing device on the loom for the creation of its design. For such a silk to be woven at all, a very sophisticated drawloom must have existed in the imperial Byzantine workshop at Zeuxippos named in the inscription of the silk.

Becker suggested that the most advanced hand draw-looms had comber boards, a device for evenly spacing the warps across the loom, but he did not discuss whether this was a device of the Byzantines or a later European development. Essentially, Becker distinguished three types of hand draw-looms: an Indian or Persian type with cross harness, a Chinese type with individually weighted draw cords, and a developed European draw-loom with comber board. D. De Jonghe divided the looms into essentially two types: those with simple heddle rods and those with necking cords and comber board. W. Endrei discerned no fewer than twelve different types of medieval looms.

D. Kuhn distinguished four forms of the Chinese hand draw-loom alone, datable before the thirteenth century. One Chinese source details the exact construction of hand draw-looms, but the account breaks off at the very point of describing the construction of the pattern-producing device, the figure harness.

Overall one must conclude that it is not possible to see a single line of development throughout. Different workshops would have required different kinds of looms for a variety of weaves. The skill of individual weavers also would have dictated the form and use of a number of looms. Small-scale workshops weaving simple patterns required looms different from large-scale workshops weaving complex and costly pieces. The surviving Byzantine silks give a clue to the stage of development of Byzantine hand draw-looms at different periods between the sixth and the twelfth century.

Byzantine Looms

Unpatterned or simply ornamented Byzantine silks, including some examples that survive in Sens, could have been woven on looms with relatively simple pattern-producing devices. However, by the seventh to eighth century, some form of advanced pattern producing device would have been necessary on Byzantine looms. There may not at first have been devices for automatic repeat of the pattern, as irregularly sized lionstrangler motifs on a Sens silk demonstrate. But to weave hunter and charioteer themes like those of surviving Byzantine silks at Aachen, sound pattern-producing devices or figure harnesses must have existed.

Unfortunately, no pieces of looms have been excavated in archaeological digs. Bricklined trenches were discovered in monks' cells at the monastery of Epiphanios in Thebes, but suggestions that these might once have contained horizontal looms with treadles have been dismissed. Documentary sources are more forthcoming. For instance, St. Theodoret of Cyrrhus in Syria, in his treatise On Providence, described a working loom: "Female hands take and spin fine threads. After stretching some threads in order on the looms, they insert the wefts.

With the rods they separate the warps, and they loosen some or tighten some attached threads, and the weft thread is then pushed across and beaten down, and so they make the cloth. . . . And . . . Who would admire how from the one color of the underlying wool or silk threads, myriad images of various animals and human forms, some hunting and some praying, and pictures of trees and many other things are woven." This passage clearly indicates shed sticks and draw loops in use on Byzantine looms in Syria by the fifth century. It is difficult to reconstruct exactly how the patterning device looked, but it is reasonable to suggest that such a loom could have been developed independently of any Chinese prototype. Chinese silks were predominantly warp faced, whereas Byzantine silks were weft faced. Chinese looms, in any event, would not have been ideally suited to Byzantine weaving techniques.

From the early beginnings of a draw-loom with figure harness system in Syria, more developed systems grew. The most advanced draw-looms may have been built under imperial patronage. The imperial Aachen Elephant silk, of the early eleventh century, demonstrates that such looms were available for the weaving of splendid silks. In modern-day India there are experiments with building hand draw-looms that help indicate the complexity of some of the advanced Byzantine hand draw-looms that existed. For example, one hand draw-loom at a Government Weavers service centre in Bangalore in India in the late 1980s was capable of 1,400 steps to produce a pattern repeat.

The Bangalore loom was more than 2 m wide and operated by a total of four weavers. Two weavers were seated at the front of the loom (on either side), and two more weavers sat facing them on a plank supported above the warp. Each of the four weavers controlled the pattern across one-quarter of the width of the loom. Other advanced hand draw-looms in India in the 1980s had pattern-producing devices suspended high above the warps, and up to six draw-boys were sometimes required to operate such devices. Such intricacies should be borne in mind when trying to envisage the Byzantine loom that wove, for example, the Aachen Elephant silk with its 1,440-step pattern repeat.

A number of eighteenth- and nineteenth-century sources depict advanced Chinese draw-looms with the weavers seated in front of the loom and with a draw-boy seated in a wooden structure high above the loom. Whether the Byzantine hand draw-looms employed this system, or whether the draw cords were led down to the side of the loom to be worked by draw-boys situated on the ground is not documented.

J.Wild and others have demonstrated how early damasks could be woven on relatively simple looms. But a significant development had occurred by around the year 1000 using some form of far more advanced Byzantine hand draw-loom. At that date, Byzantine hand draw-looms (largely built to accommodate the predominant twill weave) had to be adapted to weave lampas weaves. Monochrome lampas weave silks could be produced more cheaply than their polychrome twill counterparts. This was because in lampas the main warps as well as the binding warps were used to bind the patterns, making less labour-intensive lifting of warps necessary to produce identical patterns. Different griffin plus panther design silks were taken from the grave of Pope Clement II (died 1047). Some of these were woven as twills, others as lampases, and they demonstrate perfectly this transitional stage of development in Byzantine silk weaving. Without doubt, such advances in Byzantine weaving technology were driven by economic concerns.

Byzantine Dyes and Dyeing Techniques

The scientific study of Byzantine dyes is relatively new. To the naked eye, the extant Byzantine silks reveal that a wide color palette was in use by the twelfth century. In particular, the silks indicate that color ranges varied according to period fashions. For instance, a bright polychrome palette of reds, blues, greens, ochers, and off-whites was in vogue in the eighth to ninth century. By the tenth to the eleventh century, side by side with a still comparatively brightly colored mixed palette, monochrome tones were in demand. Single-color golden yellow, purple-blue, olive green, or cherry red Byzantine silks (either incised twills or lampas weaves) datable to the tenth to eleventh century, survive in quantity.

Early sources show initially how far Byzantine textile dyeing was dominated by concerns to build up and then protect an imperial monopoly over the use of certain murex purple dyes. Such dyes were reserved to treat specially tailored imperial garments. The Theodosian and Justinianic codes, and later the Basilics and the novels of Leo VI all strongly underline the imperial exclusivity of these murex purple dyes.

G. Steigerwald traced the growth of an imperial monopoly over murex purple from late Roman to early Byzantine times. He demonstrated that up to the period of Theodosios I (383-395), it was only the purple chlamys that was exclusively reserved for imperial use. Nevertheless, Theodosios I reinforced the edicts of his imperial predecessors Gratian (367-383) and Valentinian II (375-392) and also forbade private manufacture and use of "blatta, oxyblatta, hyacinthina, and their imitations." Not until the time of Theodosios II (408-450) were specific dress regulations encoded in "De vestibus holoveris et auratis" of the year 424.

Steigerwald convincingly argued that the latter did not provide for an imperial monopoly over all murex purples but only over murex-dyed textiles of particular imperial cut. Up to this time members of the Senate as well as private, wealthy Byzantine citizens had worn murex purples. Justinian found it necessary to repromulgate the edict of 424, which suggests that it had met with some opposition in the fifth to sixth century. A Theodosian Code edict of 333 demonstrates the existence of fraudulent dyeing practices in the imperial purple dyeing factory, and it suggests that a black market existed. A further edict of 436, repromulgated under Justinian, drew attention to illegal dyeing in the imperial purple factory of Phoenicia.

Justinian did release imperial half-blatta silks for use by wealthy females in the sixth century, but the practice did not last long. Leo VI (886-912) allowed citizens to wear clippings of imperial purple, and he castigated his predecessors for earlier denying the populace even this right. The Basilics, on the other hand, forbade illicit manufacture of murex purple on pain of death. The Book of the *Eparch of Leo VI* (911/912) divided silks into those totally forbidden for nonimperial manufacture and those to be manufactured by the private silk guilds in Constantinople, but only under the supervision of the eparch.

A wide range of variously colored purples were cited among these silks. Elsewhere some specific terms for such purples, many of which occur in the *Baggage Train account of the Book of Ceremonies*. Among the dye analyses on Byzantine silks carried out for the author in the early 1980s, it is interesting to note that one of the imperial Lion silks tested revealed a mixture of indigo and madder as constituents of its purple dye, and not murex dye. The silk appears to be one in a series of diplomatic gifts dispatched to the Latin West in conjunction with marriage negotiations. It is not unlikely that the Latins would have believed it to be a precious murex-dyed piece.

Islamic dyers used a mixture of indigo and a lichen to obtain a similar imitation purple on another silk once mistaken for a Byzantine Lion fabric. This appears to be a copy of a Byzantine imperial Lion silk sent as a diplomatic gift to the Islamic world.

Most recently, High Pressure Liquid Chromatography has been used to detect Byzantine dyes. Such tests have highlighted the use of madder and kermes as well as brazilwood dyes. Murex purple itself occurs on a fine Byzantine griffin silk now at Sitten cathedral. Murex purple has been detected also (by another process involving vatting, exposure to light, and reoxidation) on a tiny silk scrap taken from the Three Kings Shrine in the Aachen cathedral treasury.

A yellow dye was obtained from weld, and for greens, indigo and weld were mixed. A combination of weld and indigo is found on the Hungarian coronation mantle of Roger of Sicily in Vienna. The Sicilian workshop that manufactured the silk employed both Islamic and Byzantine silk workers, and it is difficult to be certain whether Byzantine or Islamic craftsmen were the dyers of the piece. Further tests are necessary to establish whether or not the off-white surviving Byzantine silks are dyed. Some may have been totally undyed and only used natural silk yarn, which itself can range in color from off-white to golden yellow.

Guild Organisation

It is impossible to decipher the guild regulations in the Book of the Eparch without reference to the actual stages of manufacture of silk yarns and subsequently of silk cloths. Only through a close working knowledge of nonmechanised silk manufacture is it possible to understand the remarkable complexity of the organisation of the nonimperial Byzantine silk industry.

The essential processes described in the Book of the Eparch relate to the work of the guilds of the Metaxopratai, the Katartarioi, and the Serikarioi. The guild regulations demonstrate the specialist division of labour not only in raw silk retailing and in silk yarn preparation but also in silk weaving. Essentially the Metaxopratai, operating in a cartel, purchased but did not work the raw silk. They could only purchase the raw silk in the capital, and they paid tax on the purchase. They had to sell the worked raw silk in approved public markets, and they were not to pass it on illegally.

Any cocoon silk would have had to be unraveled and reeled. Then it, as well as already reeled silk, would have required degumming (through boiling), and subsequently the silk would have needed to be wound into hanks. Some of the silk yarn would have been given extra twist in order for it to serve as strong warp thread. The silk from the centre of cocoons was waste silk, and this would not have been reelable; instead it would have needed to be spun.

The Katartarioi (and also the Melathrarioi and, in addition, unnamed workers hired by the Metaxopratai) variously worked the raw silk into yarn, but of these only the Katartarioi and the Melathrarioi could purchase raw silk in a cartel with the Metaxopratai. The raw silk that arrived in cocoon form, according to my interpretation of the Book of the Eparch, would have been unraveled and reeled by the unnamed workers hired by the Metaxopratai. The Katartarioi were most likely the degummers of filament silk (i.e., those who boiled off the sericin gum from the unraveled and reeled silk). The Melathrarioi, who were poorer guild members, most probably were given the task of spinning the waste silk from the centre of cocoons.

Within the workshops, too, there were many distinct skills: loom builders, pattern harness makers, weavers, draw-boys, dyers, and tailors all had to be accommodated. The Book of the Eparch described the tenth-century private silk-weaving workshops as a form of communal home to all these distinct, specialised workers. Thus the Serikarioi appear to have been an umbrella guild under which weavers, dyers, and tailors operated. The Metaxopratai sold a limited weight of raw silk to private Byzantine citizens for manufacture of private clothing in private houses.

The Katartarioi received their supplies of raw silk from the Metaxopratai (only as much as they could work), as did the Melathrarioi. The Serikarioi were also dependent on the Metaxopratai for raw silk. The Serikarioi did have the power to hire workers, but it seems unlikely that they bought anything other than unraveled and reeled silk from the Metaxopratai. Any extra workers mentioned in the Book of the Eparch were most likely to have been employed to add extra twist to the yarn. The Book of the Eparch fails to reveal to whom the Katartarioi passed on their silk. Perhaps they returned it to the Metaxopratai, or possibly they sold the degummed silk to the Serikarioi. Surviving Byzantine silks all use degummed as opposed to gummed silk yarn.

It has been suggested that the Serikarioi were multiskilled craftsmen crossing all specialties from dressing to weaving, dyeing, and tailoring the silks. This is entirely impractical. Each skill is a lifetime occupation, and different abilities are involved in each task. In addition, some occupations are too dirty to be carried out in the vicinity of weaving itself. For instance, any dyeing operation would have to be kept well away from the weaving, which had to be carried out in immaculate conditions to avoid spoiling valuable cloths. In such a factory setting, it was possible for yarn to be plied or twisted, and for the different work of weavers, dyers, and tailors to take place in quite separate workshop spaces.

Trade Names

It is particularly in regulations defining forbidden goods or kekolymena that specific trade names for fine Byzantine silks occur. Important sources for these names include the *Book of the Eparch and the Baggage* Train account appended to the Book of Ceremonies, and some precious silks also appear in the wills of provincial magnates. In many cases, trade names have no parallels from which to draw meaning outside the silk trade. On the whole, they can be best interpreted with reference to practical processes involved in their manufacture.

The last two terms refer to peach and to green-purple hues. The dimoi´rwn ojxe´wn have been interpreted to refer to redpurples. The term -pwlon is not entirely clear. Italian silks of the thirteenth to fourteenth century certainly were regulated in this manner. To ensure the correct number of warps, they were accurately threaded in the correct proportions through a reed. If the term did refer to the warps, then it would also suggest the presence of a reed on the Byzantine loom. This is not unlikely, judging by the even spacing of warp threads in extant Byzantine silks by the tenth century.

The term is less likely to apply to size of design, particularly as small intricate designs in complex weaves perhaps with brocading, for instance, could easily demand higher prices than those with large-scale, simpler, nonbrocaded patterns. In any event, cost was not so much entirely dependent on scale as on whether the designs were symmetrical or asymmetrical. Mirror-image designs, such as hunters in medallions, could be woven using the patternmaking device in a straightforward manner for the first half of the design and then simply in reverse for the mirror-imaged remaining half of the pattern.

Asymmetrical designs, on the other hand, required far more manipulation of the pattern-producing device, as there was no reversal of any part of the design entailed. The number of pattern cords required in the case of the symmetrical design was greatly reduced in comparison to that needed for its asymmetrical counterpart. A small-scale asymmetrical design could be far more costly to weave than a larger-scale symmetrical pattern. Predetermined price ranges according to pattern size, in these circumstances, would be very difficult to operate.

Most plausibly, the former terms also designated numbers and weights of warp threads employed, as discussed above in connection with the term. Other explanations including the value of the silks, degree of gold ornament, and number of loom widths involved in tailoring garments have no technical basis. Silk values depended on too many variables. Gold fabrics could have been, and since earlier times indeed were, described with terms that included the word gold. Surviving silks demonstrate that silks were not woven in narrow loom widths and that very ample looms existed. It would have been unnecessary to tailor using narrow strips. Where silks with selvedges have survived, it is possible to detect Byzantine looms up to more than 2 m wide.

The terms ajrjrJa´fia and ejrrame´na, it has been suggested, relate to garments with or without sewn panels. There is no evidence of such silks. It is more plausible to suggest that the terms mean sewn or unsewn and that they distinguish garments woven in the piece from tailored garments; that is, they served to differentiate "woven to shape" items from tailored garments. The term di´aspra probably meant two-tone white and referred to silks that relied on weave changes rather than on color contrasts for the formation of their designs (e.g., damasks and lampases).

Placing Value

The *Book of the Eparch* regulated that all silks priced above 10 nomismata were to be declared to the eparch. The Russian Primary Chronicle stated that "when the Russes enter the city, they shall not have the right to buy silk above the value of fifty bezants. Whoever purchases such silks shall exhibit them to the imperial officer, who will stamp and return them."

The price of individual silks in large part depended on the cost of the raw materials involved. Byzantine raw silk was not traded openly, and it does not appear on price lists together with other raw silks. The average price of standard-quality raw silk traded in the Islamic Mediterranean in the tenth to eleventh century was 2.5 dinars per pound (5.5 dinars per kg). This sum was sufficient to maintain a family for one month.

On certain silks, up to half the cost of production might be accounted for by the use of precious dyes. Purple dyes in particular could add to the costs heavily. From the time of the Edict of Diocletian, a special price range existed for different purples. In A.D. 300 the edict indicates that when raw silk cost 12,000 denarii per pound, murexdyed raw silk fetched an astonishing 150,000 denarii. These prices can best be appreciated through comparison with others cited in the edict. For instance, a haircut cost 2 denarii. A plain silk weaver (and also a sewer cleaner) per day received 25 denarii plus maintenance. Clearly a distinction was made in costs to allow for payment of lesser and more greatly skilled weavers producing simpler and more complex weaves and designs.

Prices for middle Byzantine precious dyes, including murex purple silks, are not available, but an incident recorded in the Cairo Geniza documentation suggests that they were still exorbitant. Cambridge University Library document Or 1081 J9 indicates the fate of a Jewish imperial dyeworks employee who inadvertently spoiled an imperial silk. His children were held for ransom while he (after being tortured almost to death) fled to Cairo to seek help from his Jewish coreligionists.

The Cairo Geniza documentation of the tenth to eleventh century provides specific details about the value of silks in the dowries of Jewish Mediterranean brides. These bridal trousseau inventories indicate the relatively high value placed upon Byzantine brocaded furnishings in particular. For instance, an undated document records one divan of unknown size, of Rumi brocade, and with two cushions and a back, at 50 dinars. Another document, dated 1156, lists a Rumi brocade divan at 40 dinars.

Overall prices varied somewhat. For instance, whereas one Rumi brocade divan without back or cushions in 1140 cost 15 dinars, at the same date, another six-section divan of Rumi brocade was listed at 40 dinars. S. D. Goitein assigned the more expensive example to a rich household and the less expensive one to a lower middle-class household. A bridal diary of the same period (ca. 1140) included "a bed cover of Rumi brocade" at 10 dinars, a "sofa of Rumi brocade (six pieces)" at 40 dinars, and a "sofa of Rumi brocade (3 pieces)" at 15 dinars.

9

Technologies in Silk Production

The conventional breeding methods such as progeny testing, exploitation of hybrid vigour, genotype ´ environment interaction coupled with utilisation of silkworm stocks that carry a translocated W chromosome provided continued success. Recent developments in transgenic silkworm technology, application of DNA markers for strain characterisation and construction of linkage maps, and understanding the genetics of viral resistance provide requisite tools that can expedite further silkworm improvement.

The domesticated silkworm, *Bombyx mori* is one of the genetically well-characterised insects next only to the fruitfly, Drosophila and has recently emerged as a lepidopteran molecular model system. The well-developed genetics of this species includes more than 400 well described mutations which have been mapped to > 200 loci, comprising 28 linkage groups or chromosomes. In addition, hundreds of geographical races and genetically improved strains are maintained in different countries where sericulture was/is in vogue. These races and strains differ not only in well characterised Mendelian traits but also in not so well-studied complex or quantitative traits such as body size, feeding duration, thermal tolerance, and disease resistance.

The monophagous nature of silkworm initially confined its rearing only to those countries where its food plant, mulberry (Morus species) is cultivated for silk production on industrial scale. However, research carried out in the early 1960s on the nutritional requirements of the silkworm led to the formulation of artificial diet, which freed silkworm rearing from the use of mulberry leaves and helped to expand the study of silkworm to academic laboratories.

Genetic Strategies

Cross-breeding Strategies and Exploitation of Hybrid Vigour

As with many agricultural crops and livestock, silkworm improvement through conventional breeding strategies followed farmers' requirements and needs of the markets, with reasonable degree of success. However, the breeding objectives followed in different countries where sericulture was/is being practised remained quite different and quite often contradictory for many of the traits. For example, in Japan, which achieved remarkable success in improving the qualitative and quantitative output of silk, the breeding objectives centered on obtaining higher silk productivity of superior quality. Silk productivity per larva improved remarkably over the years.

Conventional hybridisation and selection was and still is widely used for improving yield potential in silkworm. This strategy generates variability by hybridising the elite genotype (the breeding lines having the desired traits) with other improved varieties or local varieties, followed by selection of the desirable recombinants. The breeding objectives were achieved with precise selection starting from the initial choice of parents, meticulous progeny testing and selection and matching ecological and nutritive conditions.

To further increase the yield potential of silkworm, heterosis breeding - which exploits the increased vigour of the F_1 hybrid (improved growth rates, better crop stability by virtue of increased resistance to biotic and abiotic stresses) has been used in silkworm. Today, most sericulturists rear silkworm hybrids in almost all the countries where sericulture is in vogue.

In fact, the two big leaps involving the use of hybrid vigour were taken with the silkworm and maize, the former coming slightly earlier than the maize. The demonstration that hybrid vigour is manifested in F_1 hybrids when two genetically distinct silkworm strains are crossed was carried out by Toyama as early as in 1906. The superiority of F_1 hybrids was so much apparent that by 1919, 90% of the eggs produced were hybrid borne, reaching 100% by 1928, in Japan. The average weight of the cocoon shell, increased by selection from around 180 mg in 1804 to about 200 mg in 1910, was further enhanced thereafter to more than 400 mg in less than 20 years, when the F_1 hybrids were introduced in Japan. Another important application is the use of tetra-parental hybrids involving the crossing of two F_1 hybrids, each produced by different combinations of Japanese or Chinese silkworm stocks.

Translocated W-chromosome Stocks

Sex chromosome constitution in silkworm is ZZ in males and ZW in females. Hasimoto unequivocally demonstrated that the W chromosome nearly monopolises the power of determination of femaleness through the findings that even triploids with ZZW and tetraploids with ZZZW chromosome constitution become normal females (Table 1). The conclusive proof of evidence that W chromosome determines femaleness came when the W chromosome carrying the translocated dominant gene for larval markings from second chromosome was discovered by Tazima in an irradiated batch of eggs.

Table 1. Sex determination in polyploids, with special reference to the ratio of sex chromosomes to autosomes

Ploidy	*Number of autosomal sets*	*Number of Z-chromosomes*	*Number of W-chromosomes*	*Sex expression*
2n	AA	ZZ	-	Male
Z			W	Female
		Z	Jl.W.Z.	Female
		Z	-	Male
3n	AAA	ZZ	-	Male
		ZZZ	-	Male
		ZZ	WW	Female
		ZZ	W	Female
		Z	WW	Female
4n	AAAA	ZZZ	W Female	
		ZZ	WW	Female
		ZZZZ	-	Male
6n	AAAAAA	ZZZZ	WW	Female

When the translocated dominant gene was present, the larvae carrying the markings were invariably females, while all the larvae devoid of markings were males. This discovery not only confirmed the female determining role of the W chromosome but also led to an important sericultural application where males could be separated from females at larval stage for the purpose of making precise hybrids between two different silkworm strains.

Such translocated stocks came to be known as sex-limited strains.

After this initial discovery, strains with sex-limited cocoon or egg colour were generated by translocating the yellow cocoon colour gene from chromosome 2 and the black egg colour gene from chromosome 10 to the W chromosome. Sex-limited strains became a reality and heralded a new era in the application of fundamental genetic principles to the commercial use in sericulture. In all the sexlimited strains, whether the discrimination is at larval, egg or cocoon stage, females and males can be easily separated and little labour is required for such a task.

Besides saving labour, the sex-limited strains have an added advantage. The females generally consume more leaf and produce lesser quantity of silk per unit amount of leaf than males since some portion of nutrition is allocated for egg production. In Japan, when the sericulture was in its glory, more than 60% of the total silk production was contributed by sex-limited breeds.

Genetic Correlation of Traits

Silkworm breeders have uncovered the fact that some characters or aspects of them affect other characters. Each of these characters has its own optimal mean value which is commensurate with the adaptive fitness of the strain. The interrelationships between characters are expressed in statistical terms, as correlations, which show how one variable changes as the other changes. Positive correlations show that as breeders change the mean of one character towards the higher side, the other also goes up with it, while in the negative as the mean value of one character goes up, the value for the other character goes down.

In silkworm, most of the correlations between different characters have been worked out by selection experiments. These studies have shown that direct selection for one trait has correlation with the other quantitative characters. The correlation of some characters was found to be negative while in others it was positive. The correlations between various characters were clearly revealed by systematic studies which enabled breeders to exercise selection for various traits keeping a vital balance so that the growth, viability and silk yield are not affected. In a classical experiment by Miyahara, the silk filament length was increased from 1094 meters to a remarkable length of 2400 meters at the fiftieth generation of selection. The increase in filament length appears to have manifested with a corresponding decrease in the thickness of the silk fibre.

Viral Resistance Genetics

Silkworms are routine victims of various viral pathogens and cocoon crop losses are quite recurrent in tropical countries like India. These include cytoplamic polyhedrosis virus (CPV), nuclear polyhedrosis virus (NPV), densonucleosis virus (DNV) and infectious flacherie virus (IFV). Some silkworm strains used in tropical countries are naturally hardy and are more resistant to pathogens than the more productive temperate climate-adapted strains. Our basic knowledge about insect resistance to disease organisms mainly concerns bacterial agents; but there is a paucity of information concerning viral pathogens.

Resistance to NPV, CPV and IFV has been reported to be under the control of polygenes, except for one instance of a strain harbouring a major gene for CPV resistance, which renders breeding for resistance difficult. In contrast, a single recessive gene (nsd-1) controlling refractoriness to DNV-1 has been reported and mapped to chromosome 21. Resistance to DNV-2 has also been reported to be controlled by a major recessive gene nsd-2. Recently, molecular tags closely linked to DNV-1 resistance have been reported.

Transgenic Silkworms

Transgenic silkworms have been recently generated using a lepidopteran transposon *(piggyBac)* to integrate the foreign genes of interest into the chromosomes of germline cells. The transgenic silkworm system provides novel opportunities to improve strains of sericultural interest. Already efforts are on in the University of Lyon, France and CDFD, Hyderabad, to inculcate viral resistance by targeting baculoviral genes.

As a result of the systematic studies on selection and hybridization, high-yielding silkworm varieties were developed and sericulture vocation proved quite remunerative and cost-effective. As electronics and other industries started flourishing in Japan, the greater emphasis was laid on the reduction of manual labour in sericulture, which was very effectively achieved through introduction of many precise automated technologies in egg production, rearing, and silk reeling technologies. The success in silkworm breeding in Japan was also largely due to rearing of silkworm breeds that are native to temperate conditions. Countries like Korea, which enjoy a temperate climate, took advantage of the Japanese technologies to harness higher silk productivity.

In India where sericulture is predominant in tropical regions, silkworm breeding research is a mixed bag of success and failure. The success and spread of silkworm rearing was mostly due to the use of F_1 silkworm hybrids of exotic temperate silkworm male parental strains and native female tropical silkworm strains. The hybrid vigour in terms of survival and silk cocoon yield was achieved to an extent of 40% over the mid-parental values. However, the quality of silk achieved was a trade off between the tropical and temperate parental strains involved in the hybridisation programme.

The successful introduction of F_1 hybrids of tropical female and temperate male silkworm strains was the major contributory factor for sustainable sericulture in India. However, cocoon and silk yield and quality of silk yarn remained inferior to those of temperate silkworm hybrids. The attempts to spread temperate silkworm strains throughout the sericulture belt of India resulted in extensive crop failures, especially during hot and humid seasons and under rainfed farming conditions.

Now it has become clear that the temperate strains can only be propagated profitably during favourable seasons and under irrigated farming conditions. The breeding objectives followed to generate high yielding bivoltine varieties in the breeding laboratories centered around higher body weight, higher fecundity, longer silk fibre, almost all of which had negative correlation with the survivability of silkworm. Unfortunately, such negative correlations tend to get pronounced under poor ecological and nutritive conditions. As a result, gap between genetic and realised potentials of the silkworm hybrids remained quite large. (Fig.1)

The contributory factors for this gap need to be analysed and they seem to lie very much in the silkworm breeding process and ecological and nutritional status in which farmers raise silkworms. In the years ahead, there is a need to develop silkworm varieties with higher yield potential, more importantly with higher yield stability. In the years to come, traditional breeding methods with additional refinements will continue to be used, supplemented by tools and techniques developed by recent advances in silkworm genomics, which are offering novel opportunities to facilitate the development of new silkworm varieties. Typical breeding methods practised by Japanese breeders cannot be simply repeated under Indian conditions hoping to get the kind of success they had achieved.

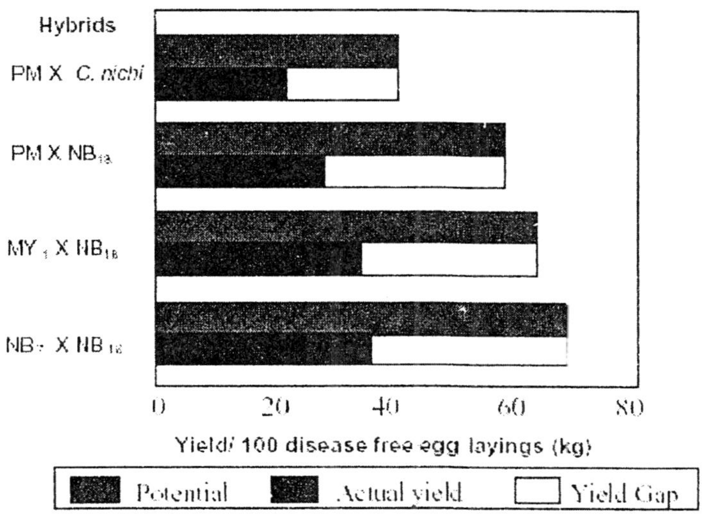

Figure 1. Yield gaps between full genetic potential and realized potential in polyvoltine × polyvoltine (PM × C. nichi), polyvoltine × bivoltine (PM × NB_{18} and MY1 × NB_{18}) and bivoltine × bivoltine (NB_7 × NB_{18}) silkworm hybrids.

Thus the important issues to be addressed are:

— There is a need to integrate the physiological, nutritional and ecological requirements of the silkworm strains under tropical conditions to the Indian silkworm breeding programmes to make them need-based.

— The need to make silkworm breeding programmes more 'scientific' given the harsh realities of 'genotype ´ environment' interaction.

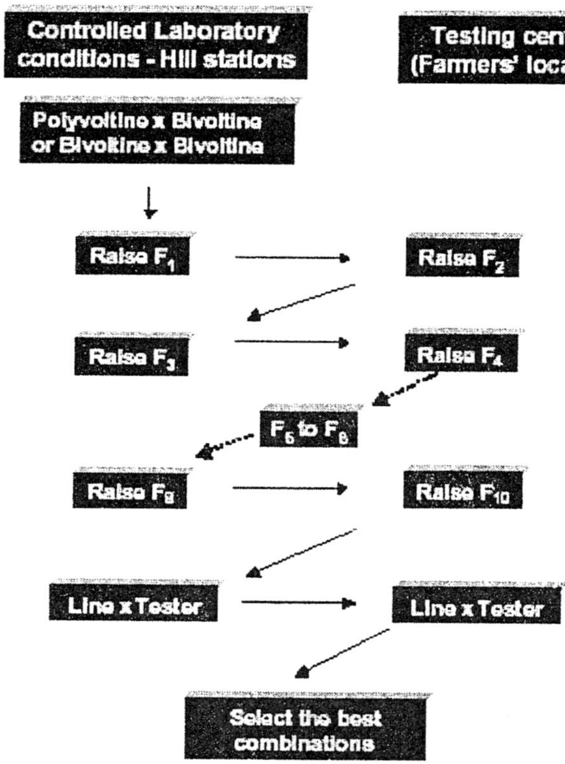

Figure 2. Shuttle breeding strategy for breeding silkworm varieties aimed at reducing yield gaps.

— Shuttle breeding approach, a concept conceived and used successfully in wheat breeding programmes by Norman Borlaug could be tried in silkworm breeding. This method allows selection of individuals adapted to diverse environments by breeding alternate generations of segregating populations at different locations. (Fig. 2)

— Adoption of well-proven genetic principles (tetraparental hybrid, introduction of translocated Wchromosome stocks for easy discrimination of sex, large scale progeny testing, screening for monogenic controlled densonucleosis virus resistance) to the silkworm improvement programmes.

— Maintenance of a broad genetic base in the early segregating populations of breeding, use of efficient selection procedures, testing and multiplication procedures, taking various genetic correlations and population size into account.

— Adoption of 'modern biotechnological tools' (such as use of DNA marker-assisted selection (MAS), selection of parental strains for cross-breeding programme based on their genetic homozygosity and genetic distance based on DNA marker evaluation, transgenic silkworm strains carrying genes for viral and protozoan disease resistance and identification of markers for QTLs) to the silkworm breeding programmes.

10

By-products Utilisation

There are theree activities involved in sericulture—production of mulberry leaves (Moriculture), rearing of silkworm and reeling and its related processing activities. In all the three sectors, certain amount of waste and by-products are generated. Different types of material wastes are generated during the production of silk goods. The by-products and wastes thus, generated during the transformation process constitute inevitable and evitable portions. The wastes cannot be recycled, reclaimed or recovered but the possibilities of reuse are many. Efforts have to be intensified for better utilisation of these by-products.

The silk industry is a labour intensive, traditional and cottage industry without much modernisation giving rise to increased amount of evitable wastes. The silk waste material constitutes considerable portion of the total cost of the final product and as such, all efforts should be made to minimise its generation in each stage of transformation process. The waste generated during reeling and further processing of silk needs more attention because of the value addition. It is also highly priced by-products and a source of income for the reeler.

Classification of By-products

Depending on the source of generation, the wastes and by-products are classified as follows:

By-products of silkworm rearing

— Silkworm litter

— *Pierced cocoons*: Moth broken cocoon shell and uzi pierced cocoons

— *Cut cocoons*: Cocoons generated in grainages

— *Floss*: Fibres surrounding the cocoons

— Double cocoons

— Black or melted, flimsy, stained and urinated cocoons

By-products of reeling and further processing of silk

— *Cooking waste*: Waste generated during cooking process

— *Reeling waste*: Waste generated during reeling process proper

— *Pelade layer-last parchment/basin refuse*: Layer of the cocoon which is unreelable

— Half reeled cocoons and unreelable cocoons

— Pupae

— *Charka waste*: Generated in charka system of reeling

The wastes generated in the course of silk winding, throwing, spinning and weaving:

— *Re-reeling waster*: Non-twisted silk waste generated during the re-reeling process

— *Twisted silk waste*: Yielded in the course of winding, throwing and weaving

— *Silk noils*: Noils are yielded in the course of silk spinning. They are used for noil spinning.

Financial Problems

The wastes and by-products generated at various stages of filamental silk yarn manufacture affects the cost of production. Excessive generation of inevitable waste due to raw material characteristics, process parameters, machineries and technology affects the economics of the industry in a considerable manner. The raw material cost (i.e. cocoon cost in raw silk and raw silk cost in fabric) is to the extent of 70 to 80 per cent of the total cost of the final product which is comparatively on the higher side.

The nature of manufacturing process in addition to the raw material characteristics, by products and wastes are generated in a considerable quantity, thus increasing the cost of production. To reduce the cost of final product, these wastes have to be effectively utilised. Number of industries have come up as ancillary industries utilising the by-products of the silk industry . However, utilisation of these wastes effectively and efficiently can reduce the cost of raw silk yarn production. Intensive efforts have to be made to find out the alternate and better use for the finished products manufactured by these ancillary industries.

Utilisation Sources

Mulberry Products

Mulberry leaves: In addition to being used mainly as feed to silkworm, the mulberry leaves can be utilised for cattle feeding, preparation of chlorophyll, soap, cosmoties, hair oil and also mulberry tea which can reduce blood pressure.

Mulberry wood (stem) : It is used as firewood and fuel and in the preparation of sports goods, fine quality papers and collection of edible fungus. The tannin extracted from stem / bark, can be used as a colouring agent.

Mulberry fruits: They are used for preparation of juice, wine, jam and jelly. The fruit juice can be used for curing sore throat, fever and diseases like dyspepsia and melancholia.

Products of Silkworm Rearing

Silkworm litter: It is used in compost making, biogas production and also in preparation of cattle and poultry feed.

The silkworm faeces, generally considered a waste and at the most used as compost in India, is being very well put to use to extract a variety of products, viz., paste Chlorophyll, Sodium Copper Chlorophyllin, Pectin, Phytol, Carotene and Triacontanol which are being used in pharmaceutical and food industries .

Paste Chlorophyll

(Treatment of silkworm faeces with alcohol) Chlorophyll A 3:1 "B" ratio Uses :

— Raw material in chemical industry

— To make many kinds of water soluble chlorophyllin

Sodium Copper Chlorophyllin

— Extracted from paste chlorophyll

— Gasoline is used as solvent

— Antibacterial action and medicinal value

Uses:

To treat stomach disorders, hepatitis, acute pancreatitis, chronic nephritis, and various leucocytopenia

Pectin

— Water soluble polysaccharide

Uses:

— Widely as food additive

— Thickner in the manufacture of candy, jelly, jam and concentrated fruit juice

— Stabiliser to make sour milk, ice creams, etc,

— Has the property of reducing blood triglycerides and blood cholesterol

Silkworm Faeces

Assuming that for every 10 kg of mulberry leaves consumed, the silkworm discharges 1 kg of faeces/litter. Around 1,40,000 tonnes of silkworm faeces are estimated to be produced in India, every year. If the silkworms faeces are sold, farmers can earn Rs. 12,00,000 for every 1000 tonnes of faeces.

It is estimated that 40 metric tonnes of paste chlorophyll can be extracted from every 1000 tonnes of faeces fetching Rs. 40,00,000. Similarly, two tonnes of sodium copper chlorophyllin can be extracted fetching Rs. 60,00,000. The left over faeces after the extraction of chlorophyll can be used to produce 100 tonnes of pectin and other products like carotene, phytol and triacontanol valued at Rs. 1,60,00,000.

Cocoons products

Cocoon consists of outer layer of floss, silk shell, pelade

layer and the pupae inside. The silk filament is continuous from top layer to the inner layer. The silk industry is interested in the shell of the cocoon which varies from 12 to 24 per cent depending on the race of the silk worm. The pupa which constitutes the major portion of the cocoon weight is inevitable by-product generated in very large quantity. The process of reeling i. e. extraction of raw silk from cocoon generates silk waste which has high economic value, as it is fully fibrous material.

Pupae

Silkworm cocoon consists of two major portions, namely, cocoon shell and pupae, the percentage of which varies from race to race. For every kg of raw silk production, approximately 8 kg of green pupae are required. Pupa consists 60 per cent of moisture by weight. Hence, every kg of raw silk will yield 3.2 kg of dry pupae.

Dried pupae which consist of 28.4 per cent oil and 51.3 per cent protein are used for the manufacture of pupal oil. The pupae from all the reeling establishments are collected, enabling the reelers to earn little extra money from an useless left over material after reeling. The dried and cleaned pupae are used for the production of oil by hexane oil extraction method.

However, only two oil extraction units are operated in the country, one situated at Mysore and another at Kallur (Andhra Pradesh) utilising only the fraction of total pupae available. The units utilising purified pupae, produce oil which is brown in colour and having a very bad odour and it finds limited use in the manufacture of industrial soap. The remaining portion of the pupae after oil extraction which is rich in protein is used as poultry feed and fish feed.

The pupal oil extracted from decayed pupae is brown in colour and emits bad smell. The price of oil is less because of these two inadequacies, even though, the oil has restricted ready market. Some of the Korean made machines are used to clean the pupae of all silk wastes automatically. This machine works-out economical for bigger units. Removing odour from the pupal oil is a possible task, but it will add to the cost of production, taking away the edge over other oils used in soap making. The last layer of the cocoon shell covering the pupae is not reeled which is extracted as basin refuse. The presence of this layer poses difficulties in separation of pupae which retards sun drying hastening the process of putrefication. The problem becomes more complicated during winter and rainy seasons due to slow sun drying.

Pupal skin

- One of the high potential commercial materials for various industries
- Available in the reeling and grainage industries as waste (260 mt in Karnataka)
- Made up of Chitin (polysaccharide), structure similar to cellulose.

Chitin

- Chitin content - 3-4 % (Dry weight of pupa)
- 2.2 % (crude chitin)
- Ash content
- 1.3 (Purified chitin)

Uses :

- *Micro-crystalline chitin*—Used as an additive to increase the loaf volume of wheat flour bread from about 4.9 to 6.0 cm^3/gm

— Post operative treatment of the intranasal operations such as conchotomy, deviatomy, polypectomy, etc,

Chitosan

— It shows potent antimicrobial activities against various bacteria and fungi.

— Chitopoly is a polynosic fibre blended with chitosan.

— It has antimicrobial activities against Staphylococcus aureus, Klebsiella pneumoniae, Aspergillus niger etc.

— To decolourise dye waste water

— Useful as a biocompatible membrane to check bleeding in major surgeries

Basin Refuse

The last layer of the cocoon attached to pupa cannot be reeled. It is a source of waste generation. However, the quantity is not considerable. Pupa along with basin refuse is collected by pupal merchant and is separated in the outskirts of the village manually. The delay in separation of the two deteriorates the quality of basin refuse waste. The presence of pupal skin and pupal fragments make the opening and cleaning of waste as difficult task. Because of the above inadequacies, the wastes are not fully utilised for the manufacture of spun silk yarn.

Basin refuse waste does not have ready market. In sericulturally advanced countries, 'Bisu machines' are installed to separate pupae from the last layer of the cocoon, hence increasing the cleanliness of the waste. The investment will be economical only in medium and large scale reeling operations. Efforts should be intensified to evolve better utilisation of this waste in the manufacture of products having higher economic value.

Cut and Pierced Cocoons

Considerable quantity of cut and pierced cocoons are generated in the process of silkworm seed production. These cocoons cannot be reeled, even though it is 100 per cent fibrous material. Cut and pierced cocoons are utilised for the production of hand spun yarn (matka) using takli. The production activity of takli yarn in concentrated in Malda, West Bengal. It utilises the cocoons purchased allover India. Pierced and cut cocoons are soaked in cold water overnight and spun with spindle takli. The resultant thread is known as matka. Spinning is done as a leisure time work in the middle of house-hold duties of the women folk, mostly from poor families. Cocoons are distributed to rural women through co-operative societies and the yarns are collected back giving the labour cost. The labour cost given per kg of hand spun yarn depends on quality and yield percentage.

The agents collect the yarn and send it to Bhagalpur weavers. The woven matka fabric is finished and exported from Calcutta Centre. The advantages with respect to West Bengal state are (a) low labour cost, (b) more number of co-operative societies working on this type of activity (c) readily available market from matka yarn(d) production of special varieties of fabrics in and around Bhagalpur on handloorn and (e) export of matka varieties of fabric from Calcutta Centre.

Double Cocoons

Double cocoons are generated during silkworm rearing when two worms join together and spin single common cocoon. Double cocoons are sorted out from reeling lots and are purchased by dupion reelers. The difficulty in uniform unwinding has necessitated to go for coarse denier in the range of 100 to 150 denier, producing coarse, thick, uneven yarn. Double cocoons are reeled either on country charkha or cottage basin to produce dupion yarn. To some extent, inferior and rejected cocoons are mixed with double cocoons to produce coarser varieties of dupion silk. Indian dupion silk fabrics have high export potential.

Cooker's Waste

Cooker's waste is a coarser variety of waste produced during cooking containing higher sericin content. Reelers waste is a better variety of waste produced during reeling, which contains lower sericin content. Hence it is considered as a better grade waste. The yield of yarn from this waste is on the higher side. These wastes are classified into different groups depending on the source i.e. reeling centres, type of contrivances (Charkha, Filature), etc. The price of the silk waste depends on the quality, source and region of its origin. The silk waste generated is rich in fibre content with good fibre properties, hence should be better utilised to produce usable products from the point of view of economy. To process bulk quantities of different types of waste, for the production of different varieties and grades (quality) of yarns, spun silk mills are established both in public and private sectors.

The waste silk cannot be converted to lustrous silk yarns without the process of removal of sericin called degumming which is normally carried out by using boiling soap/soda for a period of 1 to 1.30 h. During this process, soil, stains, oil and fats sticking to the waste material will also be removed. Productivity, quality and yield of yarns depend to a large extent on the degumming process. In addition to normal soap/soda method of degumming, enzymatic method is also used to a limited extent. There are certain advantages of this method pertaining to the quality of spun yarn, yield, lustre, etc.

Material after degumming is washed, hydro extracted and dried. During degumming process, small quantity of bleaching agents are added to the bath to get more whiteness. The degummed silk waste is now suitable for further processing for the production of machine spun yarn. The machine spun silk yarns are used either in warp or weft or in both directions. They are used for the production of suitings and shirting from fine qualities of yarn. The coarser types of spun silk yarns are used for embroidered bed covers, curtains, table covers, inferior sort of shirting's, kurtas, pyjamas, etc. Some types of spun silk fabrics are also produced in combination of raw silk yarns in weft directions.

Production of Noil Yarns

The first quality of waste obtained during the process of spun silk yarn manufacture is utilised for the production of coarse varieties of yarns, which is known as 'Noil yarn'. The sequence of process for the production of noil yarn is similar to that of waste cotton system (condenser system). Coarse spun silk yarn and noil yarns are used for the manufacture of carpets.

Twisting I Thrower's Waste

The waste silk obtained during the process of winding, twisting and doubling of raw silk, preparatory processes to weaving constitutes small quantity. Some quantity of this waste is utilised for the manufacture of coarse hand spun yarns, being used for the production of thick fabrics. Coloured wastes, i.e. rejects of dyed yarns obtained at weaving stage is used to a limited extent for the production of coarse hand spun yarns which are utilised for manufacturing carpets.

Therapeutic Values of Mulberry

Mulberry contains vitamins A (Retinol), B (Thiamine- B 1, Riboflavin-B2, Niacin-B3,Pantothenic acid-B5, Pyridoxine-B6, Biotin-B8, Folic acid-B9, Choline and inositol), C (Ascorbic acid), D and E (Tocopherol) which possess various healing and therapeutic properties. "Deoxynojirimycin" (DNJ), an alkaloid extracted from the roots of Morus nigra acts as hormone inhibitior and it blocks the pathway of infection of HIV (AIDS) to some new cells through inhibiting glycosidases at the extra cellular contact point (Suryanarayana, 2002).

Cultivation of Mulberry

Cultivation of mulberry is the most important factor in the rearing of silkworm because silkworms are monophagous insects which grow by feeding only on mulberry leaves. Moreover if we look into the economics of cocoon production mulberry leaves alone accounts for about 60%. Therefore, sericulture cannot be carried out without proper cultivation of mulberry and in planning the rearing of silkworms, the production capacity of mulberry is considered as the controlling factor. The improvement in the yield and quality of mulberry and in turn cocoon can be accomplished through application of manures and fertilisers.

In any cultivation system large quantities of nutrient elements taken from the soil are harvested in the form of biomass. This is more true in the case of mulberry leaf production. Soil nutrient deficiency is of greater relevance in mulberry in view of its feed value and sensitive nature of silkworm to leaf nutrient quality. This necessitates the supply of mineral nutrient to the soil for maintenance of soil fertility. Soil fertility may be defined as the ability of the soil to provide all the essential plant nutrients in available form and in a suitable balance. Soil fertility so far has been considered in a restricted sense as physiochemical phenomenon or as an index of available nutrients. The basic principle of soil fertility is to sustain cocoon productivity and improve farmers income through judicious and efficient use of fertilisers and organic manures.

Nutrients in the Plant Growth

Sixteen nutrient elements are recognised as being essential to all plants for their normal growth and development.

— Carbon

— Nitrogen

— Calcium

— Iron

— Hydrogen

— Phosphorus

— Magnesium

— Manganese

— Oxygen

— Potassium

— Sulphur Boron

— Copper

— Zinc

— Molybdenu

— Cobalt

These nutrients are called macro, secondary and micro nutrients only based on the quantity required by the plant for its normal growth and not based on their essentiality. The plant cannot complete its life cycle with absence of even anyone of these sixteen elements.

Macronutrients

Nitrogen (N)

- Important constituent of chlorophyll, protoplasm, porline and nucleic acid.
- Increase growth and development of all living tissue
- Improves the quality of leafy vegetable and fodders

Phosphorus(P)

- A constituent of phosphatides like nucleic acids, proteins, phospholipids and co- enzymes
- Constituent of certain amino acids
- Necessary for cell division and root development,
- Necessary for meristematic growth, seed and fruit development stimulates flowering

Potassium (K)

- As an activator of enzymes involved in protein and carbohydrate metabolism
- Helps in carbohydrate translocation, synthesis of protein, stomatal opening and pH control
- Helps to resist cold and other adverse condition and disease,

— Induce plumpness of grains and improve the quality of fruits and vegetables

Secondary Nutrients

Calcium (Ca)

— Constituents of cell all necessary for normal mitosis

— Maintenance of chromosome structure

— Activator of enzymes and helps in translocation of carbohydrates

— Acts as a detoxifying agent by neutralising organic acids in plants.

Magnesium (Mg)

— Key element in chlorophyll without which photosynthesis would not occur.

— Promotes uptake and translocation of phosphorous

— Helps in movement of sugars within plants

Sulphur (S)

— Constituent of sulphur -bearing amino acids

— Helps in stabilising protein structure

— Aids in the synthesis of oils and formation of chlorophyll

Micronutrients

Zinc (Zn)

— Involved in the biosynthesis of plant hormone, indole acetic acid (IAA)

— Plays a role in nucleic acid and protein synthesis

— Helps in the utilisation of phosphorus and nitrogen in plants

Copper (Cu)

— Promotes formation of vitamin A in plants.

— Components of many enzymes

Iron (Fe)

— Necessary for the synthesis and maintenance of chlorophyll in plants

— Essential component of many enzymes

— Plays an essential role in nucleic acid metabolism

Manganese (Mn)

— Acts as catalyst in enzymatic physiological reactions in plants

— Involved in the oxidation of carbohydrate to Co_2 and H_2O

— Involved in the synthesis of chlorophyll

Boron (B)

— Essential for protein synthesis

— Changes the activities of certain enzymes

— Facilitate carbohydrate transport

— Associated with the uptake of calcium and its utilisation by plants

— Regulates potassium I calcium ratio in plants

Molybdenum (Ma)

— Associated with nitrogen utilisation and in nitrogen fixation

— Constituent of nitrate reductase and nitrogenase
— Required by rhizobia for nitrogen fixation

The nutrient requirement of mulberry per hectare per year is the highest compared to any other field crops. Mulberry requires 300 kg of nitrogen, 120 kg of phosphorus and 120 kg of potassium per hectare per year. Nearly 30-40% of total cost of production cocoon goes to fertilisers and manures alone. Increasing the productivity under cost effective system is the prime objective of crop production. Higher cost of ferliliser coupled with high labour input for its split application especially under irrigated condition warrant adoption of alternative strategies to reduce the cost of production. A viable alternative is Integrated Nutrient Management (INM). This may help a long way in solving this problem. The concept of INM is to

— Increase the efficiency of absorption and utilisation of added ferlilisers
— Collection, conservation and utilisation of organic waste materials and
— To employ the bio ferlilisers Nitrogen management Among the three major nutrients nitrogen is the first major nutrient.

The high yielding mulberry varieties will require more of nitrogen for more biomass production and it responds readily to the application of inorganic nitrogenous ferlilisers. Higher requirement of nitrogenous ferlilisers in mulberry is leading to increased cost of cultivation, as ferlilisers are not only in shorl supply but also expensive. Moreover it needs greater attention in view of its low soil availability (0.01 to 0.50%), too readily soluble and lost nature. Split application of all the three major nutrients is one way of increasing the ferliliser use efficiency. One time application will lead to greater loss of nutrients especially nitrogen. Before it is used by the plants it will undergo nitrification volatilisation and even leaching and denitrification under excessive irrigated condition.

Split Application of Nutrients to Mulberry

Application	Time	Nutrients (kg/ha)	$P_2O_5K_20$
1st application	60	60	60
2nd application	60	–	–
3rd application	60	60	60
4th application	60	–	–
5th application	60	–	–

By the above split application the nutrient use efficiency could be increased, more over the nutrients will be made available to mulberry plants when it is required for its growth and development. The normal course of absorption of mineral nutrient elements is through roots. The plants can also absorb limited quantities of nutrients through the leaf which could be provided in the form of spray solution. Supplementation of nutrients through spray solution applied on the foliage is called foliar spray. As mid term correction foliar spray of nutrients can be resorted. To this kind of practices are more so followed for micro nutrients. Foliar supplementation is done when quick response or recovery from expressed nutrient deficiency symptom is needed. In problem soils, where soil application will not be of useful, foliar spray of nutrient will increase the fertiliser use efficiency.

Foliar spray can produce faster recovery compared to soil application. At the same time, care must be taken to monitor the toxicity level of nutrient solutions as this may produce scorching effect on foliage. Different nutrients along with growth promoters and plant protection chemicals can also be used in compatible combination. But only a limited quantity of nutrient alone can be supplied to plants.

Bio-fertilisers

Bio-fertilisers (Microbial inoculants) are the products containing living cells of different types of micro-organisms which have an ability to synthesise or mobilise nutritionally important elements from non-usable to usable form through biological process. Although the advent of this phenomenon is as old as a century , the need of its commercial exploitation was not felt in traditional agriculture. The bio-fertilisers have also the ability to produce growth promoting substances like indole acetic acid (IAA), Gibberellines (GA), B vitamins and antifungal substances which favour better growth of plants. In recent years, biofertilisers have emerges as an important component of INM and hold a promise to improve the crop yields and nutrient supplies.

There are different types of bio-fertiliser micro-organisms:

— Nutrient transformation

— Source of raw material

— Nitrogen fixing micro-organisms

— Atmospheric

Phosphorus Management

Phosphorus is one of the major essential nutrient for healthy growth of plants. Except for nitrogen and potassium phosphorus is required in greater amounts by plants than any other essential nutrients. The total phosphorus content in Indian soils range from 0.006 to

0.5%. Although the amount of total phosphorus in an average mineral soil is more in comparison to nitrogen, its availability is scarce. Most of this soil phosphorus is tied chemically in a form that would not be available to the crop in a single growing season.

Even applied phosphorus is converted into unavailable forms. The conversion of soluble phosphates into insoluble form is governed by many factors, mainly by soil reaction (pH). In acidic soils, phosphorus gets fixed by iron, aluminum and manganese which are abundant in such soils. In alkaline and saline soils phosphates gels precipitated with calcium as calcium phosphate, resulting in its non-availability. The most favourable pH range of soil for phosphorus availability to mulberry plant is 6.5 to 7.5. Because of fixation problem, the efficiency of applied phosphatic fertilisers ranges from 15-20%.

Split application of 120 kg of phosphorus in 5 split doses i.e. 24 kg phosphorus per crop under irrigated condition resulted in an increase in mulberry leaf yield over recommended dose of 120 kg phosphorus in two split doses. Selection of proper phosphorus fertiliser also helps in reducing fixation. Using water soluble phosphate in neutral to alkaline soil helps to reduce fixation. By maintaining the level of organic matter, expressed in terms of organic carbon content between 0.6 to 0.75%, also the problem of fixation can be reduced.

Bio-fertilisers like phosphorus solubilising bacteria phosphobacteria can be very effectively used. These bacteria when employed in sufficient quantity produce organic acids which solubilise the unavailable form of phosphates to available form. When these bio-fertilisers are used the source of phosphorus can even be of citrate soluble. Rock phosphates are good examples of citrate soluble form. The micro organisms react on these and made the unavailable form to available form. These phosphobacteria or recommended at the rate of 10 kg/ha/yr. This should be applied in two equal splits of 5 kg along with FYM after pruning.

Vesicular-arbuscular mycorrhiza (VAM) is a fungal bio-fertiliser which, when applied colonies on the root and help to absorb the nutrients by plants especially phosphorus. This bio- fertiliser is recommended @ 100 glsq.m of nursery. This should be applied in the nursery when the plants are transplanted in the main field the root colony of VAM is also carried through. It is possible to reduce the phosphatic fertilisation to a great extent in mulberry cultivation by introducing an efficient strain of VA mycorrhiza in the rhizosphere of mulberry plants. Besides this it may also be possible to use this beneficial effect of symbiosis to combat soil borne root infecting pathogens and root nematodes.

Organic Manures

Organic manures are derived from plant, animal and human residues. Among them are bulky organic manures like farm yard manure, compost and green manure etc. Bulky organic manures contain low amounts of nutrients and their main value lies in the supply of organic matter to the soil. Unless applied in large quantities, they do not contribute much to the nutrient supply to plants. Nevertheless the organic matter added in the form of manures performs certain other essential functions. It promotes microbial activity in the soil and improves its structure, aeration and water holding capacity which in turn improves the soils capability to respond to inputs like seeds, fertilisers etc. It also supplies micronutrients and helps to make the phosphate in the soil more available to plants.

Farm Yard Manure (FYM)

As cattle are largely used in Indian agriculture, FYM is the most important among the bulky organic manures. It is a decomposed mixture of the dung and urine of cattle with the straw and litter used as bedding material and residues from the fodder fed to the cattle. Care is necessary in properly conserving the material and converting it into good quality manure. Serious losses of nutrients take place if the manure is exposed to sun and rain. FYM is good for all crops. It should be applied 4 to 6 weeks before the sowing of the crop depending on the prevailing weather conditions. It should be spread evenly over the field and mixed with the soil without delay to avoid long exposure to the weather.

Compost

Large quantities of waste materials are available as vegetable refuse and litter on farms and as vegetable and animals wastes in urban areas. These can be converted into useful compost manures by conserving and subjecting them to a controlled process of decomposition.

Green Manuring

Green manuring is another method of adding organic matter to the soil. A leafy crop, usually a legume, grown either in the field itself or cut and brought from elsewhere is burried ill the soil when the crop is about to flower. The crops usually grown for green manuring are

— *Sunnhemp*: Crotalaria juncea

— *Daincha* : Sesbania aculeata

— *Bersem*: Trifolium alexandrinum

— *Cowpea* : Vigna sinensis

Leaves of bushes and trees are also used for green manuring. Green manure crops car also be grown on bunds and waste lands. The green manure crop supplies organic matter as well as additional nitrogen particular) if it is leguminous crop. On an average, a leguminous crop producing 10 tones of green manure per hectare can add about 35 kg of nitrogen when ploughed under. Green manuring is effective only when there is sufficient water supply both for growing the crop and for providing sufficient moisture for its decomposition in the soil.

11

Potentials of Sericulture Industry

In developing countries, like India, agriculture and agro-based industries play a vital role in the improvement of rural economy. The limited availability of land, the limited cash returns, and agriculture being confined to one or two seasons in the year, have made villages to look for supporting rural industries, such as sericulture. Agriculture and sericulture are adopted simultaneously by the agriculturists in regions where the ecological conditions are favourable.

In India all the four major commercially known varieties of natural silk namely mulberry, tasar, eri and muga are produced. Silk other than mulberry silk are known as wild silk. India ranks third among the mulberry producing nations and accounts for about 13 per cent of global production. In tasar silk production India stands second next to china. India has the world monopoly for the fabulously famed golden yellowsilk, the muga, produced in Assam.

In India, over three million people are employed in various fields of sericulture. It is a cottage industry and provides ample work for the womenfolk in the rural areas in rearing silk worms, while the male members work in the fields. Recently the enforcing of new ideas by research institutions both in mulberry cultivation and silk-worm-handling among sericulturists, the industry is now practised as a main profession and as a major cash crop, of the country. Five varieties of silk worms are reared in India for producing this natural fibre. *Bombyx mori*, the silk worm, feeds on the leaves of *Morus* to produce the best quality of fibre among the different varieties of silk produced in the country.

Antherea assama is confined to only Brahmaputra Valley of India in the world. It produces the famous *mugasilk*. *Tasar* silk is a product of *Antherea mylitta*, which feeds on *Terminalia tomentosa* grown in the thick jungles of Bihar, Madhya Pradesh and Orissa. The recent introduction of *Antherea royeli* and *Antherea perniyi* has enabled the country to produce the oak tasar silk, *Phylosamia ricini*, the eri silkworm, which feeds on *Ricinus communis*, is raised in Assam and Orissa commercially. Of the total production of 2,969 tonnes of silk in India, as much as 2,445 tonnes is produced by the mulberry silkworms, *Bombyx mori*.

Table 1. Raw Silk Production in India

State	*Production (in tonnes) Mulberry*	*Production (in tonnes) Tasar*	*Production (in tonnes) Eri*	*Production (in tonnes) Muga*
Andhra Pradesh	—	1	—	—
Assam	7	—	87	41
Bihar —	234	3	—	
Himachal Pradesh	2	—	—	—
Jammu and Kashmir	68	—	—	—
Karnataka	2,036	—	—	—
Madhya Pradesh	1	117	—	—

Contd....

Contd....

Maharashtra	—	1	—	—
Meghalaya	1	—	—	—
Orissa	—	19	—	—
Punjab	1	—	—	—
Tamil Nadu	9	—	—	—
Uttar Pradesh	4	—	—	—
West Bengal	316	21	—	—
Tripura	—	—	1	—
Total	2,445	393	91	41

Mulberry silk is produced extensively in the states of Karnataka, West Bengal and Jammu and Kashmir. About 85 per cent of the country's production is contributed by the Karnataka state by rearing multivoltine hybrids of silkworm and this activity enables the sericulturists to harvest five to six crops a year. Jammu and Kashmir, owing to its salubrious climate during autumn and spring, is producing silk by rearing univoltine silkworms. Other states, namely, Andhra Pradesh, Assam, Tamil Nadu, Uttar Pradesh, Himachal Pradesh and Punjab, contribute roughly 1.8 percent to the total production of mulberry silk in India. The State wise silk production in the country is given in Table 1.

Tasar silkworms are reared traditionally by the tribal people of Madhya Pradesh, Bihar, Orissa. These 3 states mainly contribute to the production of tasar silk in the country. The recent rearing of *Antherea royeli* and *Antherea perniyi* has enabled the country to produce the oak tasar silk in the sub-Himalayan belt and in Manipur. *Muga* silk is grown exclusively in Assam and it is still considered to be a ceremonial dress by the local population. Assam produces as much as 90% of eri silk in the country by rearing eri silkworms on castor leaves.

Mulberry Silk

The systematic cultivation of mulberry, the food plant of *Bombyx mori*, is the first step in the production of mulberry silk. The total area under mulberry in India is 1,20,567 hectares, of which only 28,781 hectares is irrigated Whereas, mulberry is raised as a bush plantation in Karnataka and West Bengal, it is grown as trees in Jammu and Kashmir. Four Indian species of *Morus*, namely *Morus alba, M.indica, M. serrata and M. laevigata,* are raised as main food plants of silkworms. The propagation of mulberry in Karnataka and west Bengal is through vegetative propagation, whereas root-grafting and bud-grafting are practised in the hilly areas of Jammu and Kashmir and Uttar Pradesh.

Considering the ecological conditions, such as rainfall and the nature of soil, different systems of plantations for raising mulberry are practised in India. In the rain-fed areas, mulberry is planted at a distance of 7.6*7.6 cm whereas under irrigation the spacing of 5 cm* 5cm and the row system (4 cm between rows and 15 cm between plants) are followed. In West Bengal, a modified system of row plantation, popularly called 'Malda System' is practised even under rainfed conditions, because of the rich soil and heavy rainfall. Recently, tree plantation have ben introduced into Jammu and Kashmir.

The cost of producing mulberry has a direct impact on the cost of producing cocoons, as nearly 605 of the total cost of production of cocoons goes to the production of mulberry leaves. Experiments conducted in the moriculture divisions of the research institutions to evolve new varieties of mulberry and improved methods of cultivation have shown that over 30,000 kg of quality leaf can be produced per annum at competitive costs against 15,000 kg by adopting the traditional methods under irrigation. Many high-yielding varieties have been introduced into the country, and they not only double the leaf yield, but also maintain the succulence of the leaves, a factor which is very important under tropical conditions.

Silkworm-Rearing

In India (Karnataka state), where the temperature ranges from 16-31 degree centigrade, enjoys favourable climatic conditions for rearing the silkworm *Bombyx mori* throughout the year, whereas in West Bengal, the multivoltine silk-worm rearing is practised even under adverse conditions of temperature. In Jammu and Kashmir, the rearing of silkworms is practised once a year during May-June. Of a life span of 50 days of *B.mori,* the egg stage lasts 10 days, the larval stage lasts longest-25-30 days. The pupa stage takes 10 days.

The silkworm pass 4 moults during growth. The total quality of leaf required to raise a unit of 40,000 larvae (100 dfls) is 500-600 kg by using the traditional methods. The silkworms consume as much as 95% of the food after the third and fourth moults. At the end of the larval duration, the silkworm emits silk from its mouth and constructs a cocoon on a scaffolding. The cocoons produced by hybrids are used for extracting the silk fibre. For preparing the hybrid silkworm eggs, pure races of silkworms are raised in separate areas.

The cocoons are preserved properly in the egg-producing factories, popularly called silkworm grainages. The moths are allowed to emerge. The selected combination of moths are allowed to copulate for 4 hrs. The male after the first copulation is either rejected or used for copulating with another female. Later, the female is consigned to a dark plastic 'cellule'. She lays about 400 eggs in 24 hours. At the end of 24 hours, the female is crushed and examined for hereditary diseases. Only certified disease-free hybrid eggs are reared for industrial silk production.

Considering the various factors, such as the place of origin, voltinism, the colour of the cocoons, the larval markings, the colour, shape and size of cocoons, the silkworms are classified into different breeds. The multivoltine races and their hybrids are reared in West Bengal and Karnataka. Owing to the non-diapausing of the eggs and the copious growth of the mulberry, these breeds can be reared six times and the cocoons are harvested after 2 months. These multivoltine hybrids have a short larval duration and are generally poor yielders of cocoons.

Bivoltines, because of the interruption of diapause, can have only 2 life-cycles a year. However, its possible to break the diapause artificially by treating 1-day old eggs in Hydrochloric acid of 1.064 specific gravity at 46.1 degree centigrade. By adopting this technique, it is possible to take more than 4 crops in a year. Univoltines and bivoltines require more leaves than multivoltines. However, the yield and quality of the cocoons are superior to those of multivoltines. The multivoltine silkworms yield about 25 kg of cocoons per 40,000 eggs reared, whereas the bivoltines yield 40 kgs. The average annual yield of cocoons in India is as low as 150 kg under rainfed conditions, and under irrigated conditions, it is about 400 kg.

Reeling

The cocoons are cooked in hot water and the silk fibre is unwound from the cocoons. This process is called 'reeling'. The silk consists of two proteins, the inner core of fibroin and an outer cover of gum sericin. During reeling, the cocoons are processed in hot water at 95-97 degree centigrade for 10-15 minutes. This process is called cooking. This cooking will enable the sericin portion to get softened and make unwinding easy without breaks. The cocoons after cooking are reeled in hot water in different types of machines.

In India, 61 percent of the silk amounting to 1,320 tonnes is reeled on the country-type *charka* numbering 6,656. The silk produced with the country charka is of very poor quality, as the thread is not uniform, as it carries many slubs etc. The improved cottage-type basins have been introduced recently into India. Provision for button-holes and a proper croisure system to maintain the thickness of the fibre, and to control the defects of neatness in the cottage basin have facilitated the production of better-quality silk. As much as 806 tonnes of silk is produced by 4000 cottage basins in the country.

Large-scale basins organised scientifically are arranged in filatures for the extraction of superior quality fibre. The silk produced by the filatures is superior because of the low level of defects of cleanliness and uniformity in the thickness of the fibre. Only 8% of the total production of silk in India is contributed by filatures. The silk produced from the multivoltine races of silkworms is poor in quality and is known to have greater defects, such as lousiness, and defects in neatness and cleanliness and is of very poor quality in light of the international D grade.

The silk produced by the bivoltine races of silkworms possesses superior neatness and cleanliness, is without lousiness and has high tensile strength and stands to the international A grade. The new technology of handling silkworms in the country has shown that bivoltine silkworms, producing the international quality of silk, can be produced throughout the year in the Karnataka state in India. It has been shown that, on an average, 30-35 kg of cocoons, yielding 3-4 kg of high grade silk can be produced by rearing 40,000 eggs of bivoltines, as compared with 25 kg of cocoons and 1.4 kg of low-quality fibre from multivoltines.

Bivoltine Cocoon Harvest

Research observations have shown that the cocoon production can be increased significantly at economic costs. It is programmed to produce 800 tonnes of bivoltine silk in Karnataka alone by 1980 and increase silk production from mulberry in India to 3,500 tonnes.

Non-mulberry silk industry

Tasar culture: Three species of *Antherea* are used for the extraction of tasar silk in India. They are *Antherea mylitta*, *A. perniyi* and *A. royeli*. In the case of the non-mulberry silk produced in India, 393 tonnes is produced from *Antherea mylitta* in Madhya Pradesh, Orissa and Bihar. This silkworm is reared on trees of *Terminalia tomentosa*, *Terminalia arjuna* found in the jungles of central and north-eastern parts of India.

The tasar silkworms also exist in nature in jungles and are collected by the tribal people during certain seasons. The first crop, usually called the seed crop, is raised during May to July, whereas the commercial crop is raised during October-November. Moths emerge from the cocoons of the November crop in June. The female moths are allowed to lay eggs after copulation in small baskets called *manias*. The eggs are collected and tied to the trees so that the hatched silkworms crawl up to reach the leaves of the food plants.

Tasar silkworms are reared wild in nature and they survive exposed to the vagaries of nature and predators. They are usually green. However, yellow, blue and white larvae are also reported. The larvae pass 4 moults before they complete their duration. At the end of the larval period, the larvae spin a ring-like structure around the twin and a long peduncle before it constructs the cocoon. The cocoons are large and brown or yellow. When the rearing is practised under cultivated conditions, the rearers protect the worms from pests and predators to harvest rich crops. The cocoons are cooked in caustic potash and reeled to extract the fibre. They are also spun for manufacturing coarse fibre.

Tasar rearing: The recent introduction of *Antherea perniyi and A. royeli* on oak in Manipur has opened up new vistas for the production of superior quality tasar fibre in India. The cocoons spun by the *Antherea perniyi* can be easily reled and superior-quality fibre can be extracted. *Antherea royeli* is available in oak jungles of the sub-Himalayan belt. The cocoons are made up of 2 layers, an outer layer of thin floss and an inner thick shell of silk. The silk can be extracted only from the inner layer. Hybrids of *A. perniyi* and *A. royeli* are being exploited commercially for the production of quality tasar silk in India. A target of 650 tonnes of silk is proposed to be produced by the end of the Fifth Five-year Plan.

Muga culture: The golden-yellow silk produced by *Antherea assama* is found only in the Brahmaputra Valley of India. This species of silkworm is semi-domesticated in that the worms which crawl down at the end of their larval period are collected by the rearers. They are allowed to spin cocoons in the rearer's houses. The worms are raised on *Som (Machilus bombycina) and soalu (Litsaea polyantha)* -trees. A single female moth of muga lays 150-200 eggs after copulating with the male for 6-8 hrs. Usually, the female is tied to a 'kharika' for laying eggs.

The larvae are reared, outdoors on trees of *som* or *soalu. The* young ones are yellowish with black markings on the body. They have the habit of crawling down the trees in groups when no leaves are left. During this time, the rearer transfers the worms to another tree. At the end of the larval period, when the worms are ready to spin the cocoons, they crawl down the tree in search of a suitable place for the construction of cocoons. Such worms are collected by the rearer and are allowed to spin the cocoons in 'jali', made of dried twigs.

The cocoon has a very weak peduncle and is golden-yellow. The cocoons are boiled in soap and soda solution and are reeled on an appliance called "bhir". It is a tradition in Assam to raise *muga* silk. The total production of *muga* silk in India is 41 tonnes and is estimated that the country can produce 120 tonnes by 1980.

Eri culture: The silk produced by *Philosamia ricini* is called eri silk. It is grown in Assam and in the eastern parts of India. The heavy rainfall and humid atmosphere in these parts are suited to eri culture. The food plants for *Philosamia ricini* is castor. The alternative food plants are sp., Plumeria Ailanthus papaya, Carica utilissima, Manihot fragans, but eri culture is mainly practised on castor. This silk worm is multivoltine and is reared indoors. The eggs are white, hatch in ten days. The hatched larvae are mounted on castor leaves in the rearing-house and are allowed to grow by periodical feeding. The worms pass four moults during its larval period of 30-32 days.

Eri silkworm is generally hardy and not easily

susceptible to diseases. At the end of the larval period, the larvae crawl in search of a suitable place among castor leaves to spin the cocoons. The cocoons are usually white. However, brick red cocoons have also been observed. The cocoons of the eri silkworm cannot be reeled, as they are made up of uneven fibres. Usually, after the emergence of the moths, the cocoons are used for producing spun-yarn. Ninety-one tonnes of eri silk is produced in the country, and it is proposed to produce 375 tonnes by the end of the Fifth Plan.

Research findings have shown that under the existing acreage of food plants of silkworms, it is possible to increase the silk production 3-fold, in addition to increasing its quality. The introduction of new technology of handling *Bombyx mori* silkworms into the major sericultural tracts of the Karnataka state has enabled the state to produce a significant quality of bivoltine silk standing to the international grades. This silk was originally imported. Its addition to the import substitution, the introduction of bivoltine silk production will enable the country to enter the export market of raw silk, opening a new directional approach to the development of sericulture in the country. It is proposed to produce over 800 tonnes of bivoltine silk by the end of the Fifth Five-Year Plan. The impact of bivoltine silkworm-rearing has already been felt by the sericulturists in Karnataka.

In addition to the higher returns of over Rs 10,000 per acre and the quality of cocoons, there is a significant improvement in the production of the existing multivoltine silk, as the bivoltine males are used for cross-breeding with the multivoltines. A significant break-through in controlling the diseases of silkworms and mulberry has ben achieved through the efforts of the research institutes in India. It has been shown that silk production in India can be increased from the present standard of 25kg per hectare to 140 kg under irrigated conditions and from 10 kg per hectare to 25 kg under rainfed conditions and the country's production can be increased 3-fold. These observations have a significant influence on the employment potential of the rural people increasing the employment potential from 30.37 lakhs of persons to 37.94 lakhs by 1980.

Recent findings relating to the rearing of tasar silkworms on oak plants in the sub-Himalayan range and in Manipur have placed the country in a new set-up of producing a significant quantity of quality tasar silk. It has also opened up new avenues for introducing and improving sericulture. It has also enhanced employment potential to the tribal and poorer sector in the hilly areas. Thus the country is poised for a white revolution leading to the production of international-quality silk, both in the mulberry and tasar sector and increasing the total silk production to 3500 tonnes providing employment potential for 37 lakhs of rural people in various fields of silk production.

Seribiotechnology

Sericulture in India is an important cottage industry based on agroforestry earning foreign exchange worth about Rs.1500 crores per annum. Presently, sericulture is practised in more than 60,000 villages providing employment to 60 lakh people who hail from the weaker section of the society and are in rural areas. Silk production has reached over 15,000 tonnes and India is the second largest silk producer (18% of world production) after China (69%). Even so, our present production falls very much short of the domestic demand.

Nearly 90% of our silk is mori silk or mulberry silk produced by the silkworm *Bombyx mori*. India has unique distinction of producing all the four varieties of silk-mulberry, eri, tasar and muga. Silk industry has a lot of social, cultural and traditional linkages in the country. Sericulture has two co-ordinate components the silkworm and host plant, the leaves of which constitute the silkworm feed. Both silkworm and host plants play cardinal roles in production of raw silk. Inspite of continuous increase in silk production in the country, the quality and the productivity is still not up to the mark as compared to China.

Realising the above, the Department of Biotechnology (DBT) in association with Central Silk Board (CSB) organised many brainstorming sessions and identified thrust areas in sericulture (both mulberry and non-mulberry) in which biotechnology can play a vital role in increasing productivity, enhancing silk quality and improvement of host plants. Accordingly, a focused programme on seribiotechnology was initiated by the DBT. So far 60 R and D projects have been supported involving various leading institutions such as IISc, Bangalore; NII, New Delhi, NBPGR, New Delhi; NCL, Pune; CDFD, Hyderabad; CCMB, Hyderabad; IIT, New Delhi; IIT, Kharagpur; MKU, Madurai and University of Delhi, Delhi alongwith various centres/institutions of Central Silk Board. Out of these 60 projects, 13 projects were jointly funded with Central Silk Board.

At CSR and TI, Mysore, an IPM package has been developed for control of uzifly at field level. The implementation of IPM package resulted a suppression of 77% in uzi infestation leading to an improvement of 5.79 Kg cocoons/100 layings. The officers from 59 cocoon markets and State Sericulture Department of Karnataka have been trained in mass multiplication and release of parasitoid—*Nesolynx thymus.* Now the parasitoid is being multiplied regularly in 20 major cocoon markets for its release in the markets and also for supplying to sericulture farmers when they visit market for cocoon transaction.

At CSR and TI, Mysore eight molecular RFLP markers linked to cocoon shell character in silkworm have been identified for the first time in the world. Cocoon shell character in silkworm is a primary determinant of the silk yield. The results have opened up prospects for utilising the identified molecular markers for successful application of directional breeding in silkworm for generating improved races with high silk yield. Apart from increasing silk productivity, efforts are being made to develop silkworm as bioreactor for producing biomolecules of commercial importance. Expression of marker proteins (*luciferase* and *green fluorescent protein*) have been successfully achieved in cell lines and larval caterpillars of silkworm (*Bombyx mori*) employing recombinant Bm NPV vector harbouring reporter genes.

An ambitious silkworm genome programme has been initiated at CDFD, Hyderabad to identify and map useful genes of silkworms using molecular techniques. This will help in breeding better silkworm and developing transgenic silkworm with desired traits in future. About 400 different types of DNA markers developed in silkworm. Three productive hybrids namely—*Swarnandhra, Hemavathy* and *Kalpatharau* of silkworm developed with increase cocoon yield and also superior quality of silk by adopting a combination of conventional breeding technique and DNA markers technology.

India is now participating in "*International Consortium*

on Lepidopteron Genomics" alongwith seven other countries in which silkworm genomics is a major component. As a part of Indian efforts, the Department has launched a programme to develop EST database on mulberry and non-mulberry (muga and tropical tasar) silkworm and also to attempt functions co-relationship of these EST with desirable characteristics in silkworm. At IIT, Kharagpur and NII, New Delhi, molecular biological studies on silk protein ((fibroin and sericin) genes of non-mulberry silkworm (eri, muga and tasar) have been carried out.

Fibroin—a major protein from eri, muga and tasar have been purified and characterised. Work is in progress to clone the whole fibroin gene from these silkworms' genomic DNA library. A new and novel type of double stranded segmented RNA cytoplasmic polyhedrosis virus infecting tasar silkworm has been identified at IIT, Kharagpur. The results of this project will lead to developing control measures against cytoplasmic viral disease affecting tasar silkworm.

A prototype of Web-enabled DNA database on mulberry has been developed for the first time in the world encompassing DNA fingerprinting studies carried out on more than 150 genotypes including 48 selected elite genotypes. Efforts have been initiated for construction of framework molecular linkage map of mulberry which will be helpful in mapping and identification of useful genes in mulberry. At Delhi University, the transient expression of marker gene (*gus)* in various explants from different commercially important genotypes of mulberry have been demonstrated. This success has opened up possibilities for the production of transgenic mulberry clones with desired traits in future.

Work has been recently initiated for developing transgenic mulberry with drought and salinity tolerance. Under post harvest technology, a process has been successfully developed for degumming of silk with fungal protease enzyme as a valuable alternative to the existing methods of degumming at IIT, New Delhi. This will lead to improvement in the quality of fabric (finished products).

The Department has initiated molecular biology studies on silk proteins and diseases of non-mulberry silkworms such as muga, eri and tropical tasar and no other research group in the world is working on these indigenous silkworm species. However, there is a need for more proactive role of the user agencies—Central Silk Board and industries in integrating technology development with production as well as effective dissemination of the products and modern knowledge so developed for the benefit of sericulture industry.

Considering that sericulture serves as a means of livelihood of over 60 lakhs poor people in India, our country can provide the leadership in the development of sericulture through application of modern tools of biotechnology. However, the Indian sericulture and silk industry have to be geared itself to accept and adopt new and novel technologies offing in future.

12

World Silk Markets

Silk has been linked with sought-after creations by the biggest names in haute couture. Yet, many admirers of spectacular garments shown on the catwalks of London, Milan, New York and Paris are unaware of the modern origins of this illustrious textile.

The fact that the raw material comes from rural areas in developing countries and transition economies is in stark contrast to the affluent environment where elegant garments of famous fashion houses are presented to a select few. Similarly, readers might wonder why ITC is involved with something as luxurious as silk in its work to help developing countries to improve their exports.

In fact, today's largest producers and suppliers of raw silk and silk yarn are in Asia, with the notable exception of Brazil. A major reason why ITC stays involved is because sericulture and silk production are labour-intensive at the village level, employing both men and women at all stages of production. In China, this sector occupies some 20 million farmers, as well as 500,000 people in the silk processing industry.

In India, sericulture is a cottage industry in 59,000 villages. As one of the most labour- intensive sectors of the Indian economy, it provides full and part-time employment to some six million people. In fact, this sector has been identified as a sector of the Indian economy with strong potential to create jobs. Silk is also environmentally friendly. Silk is produced with few chemical fertilisers and practically no insecticides. Primarily made of proteins, it is close in composition to human skin, making it extremely comfortable to wear.

Challenges in Silk Market

New competition: New synthetic fibers are ever-more sophisticated. They look and feel like silk, but are easier to care for. Viscose and polyester have taken some of silk's market share.

Changing image: While the boom for sandwashed silk is past, its introduction damaged silk's luxury image. No international promotion campaigns are addressing this issue, reflecting the lack of cohesiveness among suppliers, traders and buyers in the industry.

Regional slowdown: The fast-developing economies in Asia had been showing an increasing interest in silk products, as the standard of living rose steadily. Recent economic turmoil in southeast Asia, however, has been reflected in a decline in sales.

Dismantling production centres: Silk production centres are closing. This trend is well underway in Japan and Republic of Korea, due to the industrialisation of the two countries. Millions of families in rural areas in China, Thailand, Brazil and other countries may now face the socioeconomic choice of whether or not to continue producing silk. If farmers turn from this activity for a more lucrative type of farming, the industry cannot easily recover. Working with silkworms requires strict discipline, learned by tradition through generations.

A generic silk promotion campaign could be a solution to the challenges in the industry today. Currently, there are no concerted efforts from silk producers, converters and traders to improve the image of silk in international markets. An example of what could be done can be seen in the European Union's campaign to promote linen as fashion material. In the past, linen was mostly used for home products such as tablecloths and napkins. Thanks to a decade-long campaign, linen has taken its place among other fibres used in fashion.

A campaign should aim to reposition the image of silk, capitalise on new technologies and changing market demands and to encourage silk production. As campaigns require both coordination and investment, and may take several years before they have an effect, an effective approach may be to work through industry associations in order to craft national campaigns in the largest consumer markets.

Another alternative may be for the silk industry to work with major retailers in Europe and the United States. Ideally, an industry-wide action plan will have the most impact. Based on ITC's analysis of recent trends, generic silk promotion campaigns should contain the following elements.

The image of silk: Silk should be marketed as an environmentally friendly, luxury product.

Demand for silk blends and knitted silk: Silk blends are one way to answer the challenge from synthetic fibres. Lately, silk has been blended with other fibres, such as cotton, linen, wool and even polyester. Developing country silk producers have not yet progressed very much in this field. They need to develop their research and technology to offer competitive new products. Knitted silk products are an answer to the market demand for comfort, expressed during the sandwashed silk boom.

Consumers, especially in the United States, were attracted by these garments that were soft, comfortable, casual and easy to maintain. Knitted silk has the advantage of being a casual, yet "quality" silk product, and can fetch attractive market prices. China has made progress in this product category. After introducing silk thermal underwear for outdoor living, other types of knitted silk garments are now being exported, such as T-shirts, camisoles, polo shirts and sweaters. This trend should be followed, because it may have a reasonable growth potential for a number of silk producing developing countries.

New product categories: For instance, there is scope to expand the role of silk fabrics for interior decoration.

The role of Asia: In the medium term, the role of the Asian markets will be vital. Asians will continue to be consumers, and it is likely that growth will resume in this area. It is important to safeguard their role as producers in light of the challenges facing them today.

Recognising this, the World Bank is rehabilitating the sericulture and silk production industries in Bangladesh. Since the country has long used silk fabrics for saris (like in India), greater silk production in Bangladesh is aimed at reducing raw silk imports, as well as creating a new indigenous raw material for garment exports.

ITC Acts

For 15 years now, ITC has worked with firms, national governments and industry associations to promote silk and silk products in silk producing developing countries, particularly in Asia.

ITC has closely monitored world production and trade in silk goods, working with key industry players such as the International Silk association, the International Sericulture Commission (both in Lyon, France), and relevant organisations in developing countries, including the China National Silk Import and Export Corporation, the Central Silk Board of India, the Indian Silk Export Council, the Thai Silk Association and others.

ITC projects gave an opportunity to local silk industries, particularly in China, India, Malaysia and Thailand, to develop products and markets. ITC provided technical assistance for silk fabric dyeing and printing, design, pattern-making and grading, cutting and other silk production techniques.

Participating companies were introduced to new markets and customers through market contact missions in foreign markets. The message was always the same: move toward silk processing and finished products, in order to improve value-added, and stay competitive.

ITC also publishes a Silk Review, which surveys international trends in production and trade. First published in 1988, it has been updated six times. *Silk Review 2002* is available free to developing country readers and sold at US$ 40 for readers in developed countries. In 1998, ITC produced a film "Silk: Tradition with a Future" with the United Nations Development

Sinha, S.S., "Role of NGO in Sericuture Development".Paper in International Conference on Sericuture", Banglore, India, 1994.

Sunildutt, J., and Chole, R.R., "A Study on Adoption of Sericultural Practices by Sericulturists", *Research paper on The Indian Journal Sericulture*, Vol.41, No.1, Mysore – India, 2002.

Veda, K., Nagai, I., and Horikomi, M, "Silkworm Rearing", *Translated from Japanese*, Science Publishers Inc, New Hampshire, United States of America, 1997.

Index